A STRONG AND SUDDEN ONSLAUGHT

The Cavalry Action at Hanover, Pennsylvania

John T. Krepps

Colecraft Industries
Since 1981

Since 1981

Published by Colecraft Industries
970 Mt. Carmel Road
Orrtanna, PA 17353

Copyright © 2008 by John T. Krepps

All rights reserved. No part of this publication may be reproduced, stored in a retrieval system, or transmitted, in any form by any means, electronic, mechanical, photocopying, recording or otherwise, without prior permission of the publisher.

ISBN 978-0-9777125-7-1

For more information visit our website: **www.colecraftbooks.com**

Or contact us via e-mail at: **colecraftbooks@embarqmail.com**

First Edition

PRINTED AND BOUND IN THE UNITED STATES OF AMERICA

Cover Design by Philip M. Cole

Maps provided by David Weaver

Photos credits key:

AC = Author's Collection
BBGB = Brevet Brigadier Generals in Blue, Roger D. Hunt Collection
G & B = www.generalsandbrevets.com
GPL = Pennsylvania Room, Guthrie (Hanover) Public Library
GRMICHPL = Grand Rapids History & Special Collections, Archives, Grand Rapids Public Library, Grand Rapids, Michigan
LOC = Library of Congress
NA = National Archives

ACKNOWLEDGEMENTS

A project such as this one cannot be completed without the expertise and help of many talented people. Enough individuals were a part of this endeavor that my greatest concern is leaving someone deserving of thanks out of this list. The first credit must be given to the veterans themselves, especially Charles B. Thomas of the 5th New York Cavalry. In a postwar speech, Thomas used the adjectives "sudden" and "strong" to describe his regiment's counterattack in the center of Hanover. I believe his words also provide an accurate description for several different phases of the battle, and decided to use his account for the title of this book.

Thanks and acknowledgements must go to the staffs at the following locations for their help:

National Archives
United States Army Military History Institute
Pennsylvania State Archives
Pennsylvania State Library
Maryland State Archives
Adams County Historical Society
York County Historical Society and Heritage Trust
York County Archives
Gettysburg National Military Park Library
Gettysburg Licensed Battlefield Guide Library
Pennsylvania Room, Guthrie (Hanover) Public Library
Hanover Area Historical Society

Particular mention should be given to the following individuals: At the Adams County Historical Society, Tim Smith and Randy Miller provided much assistance over the years and Larry Bolin kindly shared his knowledge on the Schwartz Schoolhouse area. My gratitude is also extended to Scott Hartwig and John Heiser of the Gettysburg National Military Park Library. All were of great help in directing me to sources of information. I cannot give enough thanks to John McGrew and Wendy Bish McGrew of the Pennsylvania Room at the Guthrie (Hanover) Public Library. John and Wendy made several suggestions and corrections on the part of the text concerning the Hanover Borough history, and directed me to numerous accounts that I would not have found on my own. John was also kind enough to scan some historical photos from the collection at the Pennsylvania Room and copied sections of the 1860 York County map for my use.

Certain individuals were gracious in providing specific accounts. Sincere gratitude is extended to Mr. J. Marshall Neathery, Rolesville, N. C. for providing the article from the *Raleigh News and Observer*, and to Wayne Motts of the Adams County Historical Society for putting me in touch with Mr. Neathery. A very big thanks to Mark Stowe of Grand Rapids, Michigan and the David Van Dyke family of Nappanee, Indiana. Mark provided a copy of his history on Company B, 6th Michigan Cavalry, plus some fascinating letters of the soldiers of that regiment. Mark also contacted the Van Dyke family, who graciously allowed me to use the Daniel Powers letter. I am also in debt to Eloise Haven of Grand Rapids, Michigan for providing her manuscript of the letters of Allen Pease. All the above accounts were of great value to my research.

Much thanks must also go to Peter and Sharon Sheppard for all their interest and support, especially their input on the Gitt's Mill area. Because of their kindness I have been able to hike and make fascinating discoveries on tracts of land where significant troop movements took place, areas that are almost completely unknown except to a few individuals who have lived in the area. I must also give credit to Bob Resig for his interest and dedication to local Civil War history. Bob's dedication, among others, was one of the main factors in the creation of the Hanover wayside markers and walking tour, and his hard work served as an inspiration throughout my own project. Thanks also goes to Mr. G. Thomas LeGore for kindly sharing his extensive knowledge concerning the Maryland areas of Westminster and Union Mills. Concerning the text of this project, credit must be extended to Cheryl Sobun, who edited the narrative and endnote sections. Thanks to her expertise, many punctuation and spelling errers [sic] were eliminated. Any remaining mistakes in those sections are due to my failure to insert her corrections.

As a Gettysburg Licensed Battlefield Guide, I come in contact with people from all over the world with an intense interest in the Civil War. I also get to work daily with Guides and Park Rangers who have given programs on many battlefields, and whose interpretive skills are among the most impressive in the historical field. The collective level of historical knowledge among the Guides and Rangers is staggering and one of the greatest aspects of it all is that I never stop learning. Several Licensed Guides, however, need to be singled out in this instance: Sue Boardman was kind enough to provide accounts of the 147th Pennsylvania. I'll bet Sue never thought I would be able to get information about her favorite infantry regiment into a book on a cavalry engagement, but I did it. Meanwhile, Gary Kross directed my attention to the powerful description by Charles Adams of the hardships endured by cavalrymen. On several occasions, Andie Custer and Mike Phipps offered valuable insight on various aspects of the cavalry movements during the campaign. I must also thank Louise Arnold-Friend for her advice and help on using the resources at the United States Army Military History Institute. Mike Kanazawich is another who should be noted for making suggestions on parts of the text. Thanks to all of them.

Some other Guides deserve even more special mention. Larry Wallace has given programs and tours on Hanover for many years, and has an outstanding wealth of information and expertise on the battle. The time he spent in looking over my manuscript and offering insights and suggestions is much appreciated. Dave Weaver deserves tremendous thanks and applause for his art work and map skills. (With the exception of the maps for Union Township, Schwartz Schoolhouse area, and Stuart's withdrawal, the base maps for this project were hand drawn by Dave. After that point, I personally inserted troop positions, etc. Any errors on the maps are my own.) Dave's attention to detail has been apparent on several occasions, including the maps he created for Licensed Guide seminars and field programs. A huge thanks must go to another Licensed Guide, Dave Richards, for much encouragement and advice, especially on all those research trips to the National Archives. D. R. also must be credited for sharing his computer knowledge; I don't think I would have been able to complete the project without his technical expertise. But possibly my biggest thanks of all, especially concerning research, must go to Tim Smith and Jim Clouse. Tim and Jim have directed me to so many different primary accounts and sources of information over the last several years that I cannot begin to count them all. Their expertise on research methods is almost unreal, and they were always quick to share their knowledge. Both also examined my writing in detail and offered several excellent suggestions for greater historical accuracy.

Finally, I must thank my Dad, Mom, and sister for all their support over the years. This project would not have taken place without them.

CONTENTS

Prologue- SITUATION, 1863	12
Chapter One - STUART'S RIDE BEGINS	19
Chapter Two - UNION	26
Chapter Three - EARLY MORNING JUNE 30	31
Chapter Four – ENCOUNTER	39
Chapter Five - STANDOFF	56
Chapter Six – ARTILLERY	60
Chapter Seven – LEE'S SCREENING MOVEMENTS	63
Chapter Eight – THE SCHWARTZ SCHOOLHOUSE FIGHT	66
Chapter Nine - THE TWELFTH CORPS REACHES LITTLESTOWN	76
Chapter Ten - HAMPTON ARRIVES	78
Chapter Eleven - WAGON TRAIN WITHDRAWAL	84
Chapter Twelve - CHAMBLISS, HAMPTON WITHDRAW	89
Chapter Thirteen - THE NEXT CRITICAL HOURS	94
Chapter Fourteen - BACK IN HANOVER….	97
Epilogue	99

Appendix A - CONFEDERATE REQUISITIONS, CIVILIAN DAMAGE CLAIMS
Appendix B - COMPANIES OR TROOPS?
Appendix C - PENNVILLE, BUTTSTOWN, MUDVILLE, MUDTOWN, OR DRECKSTHEDDEL?
Appendix D - WHICH 1ST WAS FIRST?
Appendix E - CONFEDERATE ARTILLERY LOCATIONS
Appendix F - ATTACK ROUTE OF THE 2ND NORTH CAROLINA
Appendix G - ELIZABETH SWEITZER WALTZ
Appendix H - JOHN HOFFACKER
Appendix I - LOCATION OF THE SCHWARTZ SCHOOLHOUSE FIGHT
Appendix J - GITT'S MILL SKIRMISHING
Appendix K - WHERE WERE THE WAGONS PARKED?

PREFACE

When Maj. Gen. J. E. B. Stuart began his famed cavalry ride in late June 1863, he had no idea of the chain of events that had been set in motion. Decisions made and circumstances encountered during this movement caused his troops to be detoured and delayed substantially. As a result, he would not reestablish contact with the main body of the Army of Northern Virginia until well after the critical battle had started at Gettysburg, Pennsylvania. Stuart's absence, which deprived Gen. Robert E. Lee of accurate intelligence of Union movements, clearly had far reaching effects on the invasion of Pennsylvania. The intense debate over this expedition began even before the shooting stopped during this pivotal campaign.

On June 30, 1863, an engagement took place at Hanover, Pennsylvania, between Stuart's forces and those of Brig. Gen. Hugh Judson Kilpatrick. This chance meeting forced the Confederate cavalry farther away from possible communication with Lee and added several more hours and miles to their movements. Ultimately it was not the fighting at Hanover by itself that had the greatest impact on Stuart's movements; his problems began well before he even reached Maryland, much less Pennsylvania. But the action at Hanover was an important link in the chain of events and is deserving of study by students and scholars of the Civil War.

In the early 1900s, historian George R. Prowell conducted a number of interviews with participants and witnesses of the battle. The accounts were then used as a basis for several articles in local newspapers. These articles provide a wealth of information, and we owe a great debt to this individual for his work in gathering these valuable sources. (He had previously written a short piece on the battle, which was included in John Gibson's *History of York County, Pa*, published in 1886.) Prowell also wrote his own *History of York County, Pennsylvania* (1907), which contains a section on the clash at Hanover. His writing has formed the basis of almost everything that has been published on the fighting since that time. Two later books, *The Battle of Hanover* (1945) and *Encounter at Hanover: Prelude to Gettysburg* (1962), were largely based on his work. In fact, much of the battle narrative in these two books is actually a word-for-word reprint of Prowell's writing in his York County history.

As interest in the Civil War experienced rapid growth in the late 1900s, a massive body of work emerged, with much focus on the Gettysburg Campaign. Concerning Hanover, writers have almost invariably looked to the previously cited books as the basis of their research, resulting in what could be defined as an interpretation lockout. For close to one-hundred years, much of the information in almost every publication about Hanover can be traced to the same secondary sources. But in some cases, there is overwhelming evidence that various aspects of the battle have been misinterpreted in these earlier writings. In a few other instances, items that are presented as firsthand accounts have actually been altered substantially from their original sources. Because of this, some of what the Civil War community "knows" about this battle simply does not withstand closer scrutiny but has become accepted as "history" by sheer repetition. In several appendices and footnotes, I have attempted to trace how interpretations of various aspects of the battle have evolved over the decades. Sometimes even the statements of eyewitnesses conflict, and we are left with as many questions as answers.

One of the most frustrating aspects in interpreting the engagement is the lack of Southern officers' reports. Other than Stuart's, no Confederate reports containing information on Hanover were included in the *Official Records*; likely many did not survive the war. As a result, Confederate regimental positions for much of the battle are almost impossible to specify with any certainty; even brigade locations are sometimes difficult to pinpoint.

Other difficulties occur because of the antagonism between various North Carolina and Virginia officers. For example, Capt. William Graham of the 2nd North Carolina often expressed bitterness about the treatment his regiment received during the war. Graham felt that his men had been "forsaken" by their comrades in various actions. On the other hand, Capt. William W. Blackford and Maj. Henry B. McClellan, two of Stuart's staff members, had strong ties to Virginia troops. A reading of their accounts gives the impression that the 2nd North Carolina was the only regiment involved in the Confederate assaults at Hanover and, therefore, was the only Confederate regiment that was repulsed. There was no rebuke by either of the officers concerning the conduct of any forces. Yet by leaving out the part played by the 9th and 13th Virginia, it appears to place the burden of defeat on the 2nd North Carolina alone. This interpretation, however, is in direct contradiction with not only Stuart's Official Report, but also the writings of several other officers.

The works mentioned above are still available to analyze, but there are now many other avenues of research that were previously unavailable to historians. In this study, I have attempted to utilize primary sources, whenever possible, as the basis. A large number of the sources I have cited have never before been printed in any previous publications. While other secondary works have focused on the events in the town of Hanover itself, a careful study shows much more was taking place in the surrounding townships than has been commonly supposed. In this book, I have attempted to shed some light on several of these lesser-known aspects. Emphasis was placed on previously unpublished sources to trace some of the specific roads that Confederate troops used, particularly Gen. Fitzhugh Lee's Brigade. Very few military accounts are specific as to the route his unit traveled that morning, and his screening movement has been essentially lost to history. But by examining civilian accounts, along with one notable dispatch sent by the general himself, a more detailed picture of this movement begins to emerge. I have also given much attention to battle action that took place well away from the town. It is true that most casualties took place either on, or very near, Frederick Street, within a half-hour time period. But other skirmishes broke out sporadically over at least a few hours, as detachments made contact well away from Hanover. Several years ago, one of my colleagues on the Licensed Guide force, Tim Smith, mentioned that an entire series of engagements had taken place along the axis of the Hanover-Littlestown Road that had never been examined. Tim's statement was absolutely correct, and the more accounts that have surfaced, the more fascinating this story has become for me personally. At this point, at least five distinct actions can be identified that occurred between Hanover and Littlestown, mostly after the fighting in the streets of Hanover had already subsided.

Of particular interest to me has been the mounted fighting that occurred well west of Hanover between Lee's Brigade and the 6th, then 5th, Michigan. Until the last several years no secondary works have even mentioned these particular actions. To the best of my knowledge, the first to do so was Licensed Battlefield Guide Michael Phipps in his 1995 biography of Gen. George Armstrong Custer entitled *Come On You Wolverines*. Edward Longacre's 1997 book, *Custer and his Wolverines: The Michigan Cavalry Brigade 1861-1865,* provided more detail concerning the movements of the 6th Michigan. More recently, George A. Rummel III in his 2000 study *Cavalry On the Roads to Gettysburg: Kilpatrick at Hanover and Hunterstown*, and Eric J. Wittenberg and J. David Petruzzi in their book *Plenty of Blame to Go Around: Jeb Stuart's Controversial Ride to Gettysburg* have dealt with this phase of the fighting. All these writers deserve much credit for their historical detective work. But as I began to analyze more civilian and military accounts, it became apparent that this combat was much more extensive than has ever been realized by many military historians. In researching accounts of these soldiers, one striking aspect emerges: several of the 5th Michigan men considered the fighting of their regiment to have actually not been a part of the battle at Hanover. These men were largely correct; their accounts, and those of civilians, confirm that some action occurred that was closer to Littlestown than Hanover.

As a boy, I was able to hike and fish on land owned in 1863 by Jacob Forry, Jacob Forney, Samuel Keller, and others who sustained damages during the battle. Sadly, very little of the area would be recognizable to the cavalrymen who fought here. Stretches of Plum Creek, which were fifteen feet wide and three feet deep, are now only two feet wide and six inches deep, altered beyond recognition by landscaping and development. One has to wonder if this is truly "progress." Author Nathaniel Hawthorne once said, "People and places always have a past, and their identity dissolves unless they recognize they have a history." While understanding the need for us to go forward, it is my hope that an appreciation for our past will remain.

Order of Battle - Hanover, Pennsylvania, June 30, 1863

Third Division, Cavalry Corps, Army of the Potomac
Brig. Gen. Hugh Judson Kilpatrick

Headquarters Guard
Company A, 1st Ohio - Capt. Noah Jones
Company C, 1st Ohio - Capt. Samuel N. Stanford

First Brigade
Brig. Gen. Elon J. Farnsworth

1st West Virginia - Col. Nathaniel P. Richmond
1st Vermont - Lt. Col. Addison W. Preston (Col. Edward B. Sawyer on leave of absence.)
5th New York - Maj. John Hammond
18th Pennsylvania - Lt. Col. William P. Brinton

Second Brigade
Brig. Gen. George A. Custer

1st Michigan - Col. Charles H. Town
5th Michigan - Col. Russell A. Alger
6th Michigan - Col. George Gray
7th Michigan - Col. William D. Mann

Accompanying the Division from First Brigade, Horse Artillery
Battery M, 2nd U. S. Artillery - Lieutenant Alexander C. M. Pennington, Jr.
Battery E, 4th U. S. Artillery - Lieutenant Samuel S. Elder

Stuart's Cavalry Division, Army of Northern Virginia
Major General James Ewell Brown Stuart

W. H. F. Lee's (Chambliss's) Brigade
Col. John R. Chambliss, Jr. (Brig. Gen. William H. F. Lee wounded at Brandy Station.)

2nd North Carolina - Lt. Col. William H. F. Payne (Col. Solomon Williams killed at Brandy Station.)
9th Virginia - Col. Richard L. T. Beale
10th Virginia - Col. J. Lucius Davis
13th Virginia - Maj. Joseph Gillette (Chambliss promoted to brigade command, Lt. Col. Jefferson Phillips wounded at Brandy Station.)

Fitzhugh Lee's Brigade
Brig. Gen. Fitzhugh Lee

1st Virginia - Col. James H. Drake
2nd Virginia - Col. Thomas T. Munford
3rd Virginia - Col. Thomas H. Owen
4th Virginia - Col. Williams C. Wickham
5th Virginia - Col. Thomas L. Rosser

Hampton's Brigade
Brig. Gen. Wade Hampton

1st North Carolina - Col. Laurence S. Baker
1st South Carolina - Lt. Col. John D. Twiggs/Maj. William Walker (Col. John L. Black wounded at Upperville. Most historians have usually listed Twiggs as the commanding officer. But according to Black, Walker was in command at this time. See *The Bachelder Papers, Vol. 2*, 1243, 1270.)
2nd South Carolina - Maj. Thomas J. Lipscomb (Col. M. Calbraith Butler wounded at Brandy Station.)
Cobb's Legion - Col. Pierce M. B. Young
Jeff Davis Legion - Lt. Col. J. Frederick Waring
Phillips Legion - Lt. Col. William W. Rich

Attached to above forces on June 30, 1863, were six guns from the following two batteries:
1st Stuart Horse Artillery (Breathed's Battery) - Capt. James W. Breathed
2nd Stuart Horse Artillery (McGregor's Battery) - Capt. William M. McGregor

In the Order of Battle contained in the *Official Records* the 1st Maryland Battalion is listed with Fitzhugh Lee's Brigade. The battalion, however, was accompanying Ewell's Corps at this point. The 15th Virginia was technically in W. H. F. Lee's (Chambliss's) Brigade but was on detached duty and not present on the expedition.

"I have the honor to report that after an encounter with General Stuart's force, I have succeeded in cutting his column in two."

Gen. H. Judson Kilpatrick, U. S. A., from a dispatch describing the action at Hanover, Pennsylvania

"We cut the enemy's column in twain."

Gen. J. E. B. Stuart, C. S. A., from his official report describing the action at Hanover, Pennsylvania

PROLOGUE
Situation-1863

By the late spring of 1863 the Civil War had reached a critical juncture. A series of Confederate victories in Virginia had led up to the Battle of Chancellorsville, possibly Gen. Robert E. Lee's greatest success. Southern confidence was unsurpassed after this battle, but certain realities had become apparent to the Confederate high command. Even when Lee's Army of Northern Virginia was victorious, his forces could not continue to sustain casualties like those suffered at Chancellorsville. Confederate accomplishments in Virginia could not completely offset Union gains in the Western Theater. Federal armies had captured several hundred square miles of the Ohio, Tennessee, and Mississippi River Valleys. Union forces also controlled large stretches of the Atlantic and Gulf Coasts. This strategic situation, along with the South's smaller population and lesser industrial base compared to the North, were some of the critical factors that Confederate officials were forced to consider. Lee maintained that the South had a legitimate chance to win the war, but the longer the war dragged on, the less that chance became. His forces could not afford to stand pat, and after the victory at Chancellorsville, he believed it was in the South's strategic interest to take the war onto Northern soil.

Gen. Robert E. Lee (LOC)

After vigorous debate, Southern leaders decided to adopt the general's suggested course of action. This movement was not the first time the Army of Northern Virginia had crossed the Potomac River. In September of 1862, Lee had marched close to 40,000 men into Maryland, attempting to reach Pennsylvania. Instead, the opposing forces were drawn into the battle of Antietam/Sharpsburg, Maryland, resulting in the war's bloodiest day. Like that campaign, the issues in 1863 were momentous. Moving northward would disrupt any potential plans by Union generals for a campaign in Virginia and allow the Army of Northern Virginia to gather much needed food and supplies in Pennsylvania. Lee also had his sights set on Harrisburg, Pennsylvania, which was a major training and staging area for Federal forces. If Southern forces could destroy the Pennsylvania Railroad bridge, which crossed the Susquehanna River at Harrisburg, the railway links between the Western Theater and large East Coast cities would be severed. Psychologically, the capture of that Northern state capital would have been a huge blow to Northern morale. All these benefits would have been magnified if Lee had won a decisive battle on Pennsylvania soil. Such a victory might have broken the will of the North to continue the war and possibly led to a negotiated peace and Southern independence. Just as in the Maryland Campaign of 1862, the very fate of the United States hung in the balance.

On June 3, 1863, the leading elements of Lee's army began to leave their camps near Fredericksburg, Virginia, and move westward toward the Blue Ridge Mountains. Before long, his force, with a fighting strength of about 75,000, passed through mountain gaps and then marched northward through the Shenandoah Valley. By the final week of June, large numbers of Southern infantry had already crossed the Potomac River and entered Maryland.

Meanwhile, the Union Army of the Potomac, with about 93,000 fighting troops under command of Maj. Gen. Joseph Hooker, started to march in response to Confederate movements. The Union high command did not initially know Lee's intentions, but as greater numbers of Confederates moved closer to the Potomac, it soon became apparent that an actual invasion of the North was underway. Word spread like wildfire through the Northern newspapers and by word of mouth. As with any event of great excitement, much of the information passed along was unreliable and sometimes completely false. But this much was certain: Lee's famed army was headed toward Pennsylvania.

A PROSPEROUS TOWN

The history and geography of an area are often deeply connected. Hanover, Pennsylvania is no different in that regard. The center of town is located in southwestern York County, about six miles north of the Pennsylvania/Maryland state line, and is situated on a relatively level plain. Much of the vicinity, however, is overlooked by a few significant terrain features, which would have appeared much more prominent in 1863 than they do today. The Pigeon Hills, less than three miles to the north and northeast of the center of Hanover, towers above the valley in which the borough is situated. To the south, another long line of higher ground overlooks the area, extending roughly southwest to northeast. From a distance, this elevation appears to be a single ridgeline, but it is actually a large expanse of various hills, ridges, and creek valleys. To the east and west of town lies mostly gently rolling farmland.

By the 1860s, Hanover was a thriving community of more than 1,600 citizens.[1] Five major roads radiated from the center of town, and most residents could look out their front windows upon the social and commercial activity taking place on these streets. A look out of many back windows revealed a much more pastoral setting of open land, rich with crops and dotted by an occasional barn or farmhouse. Like many towns of the era, Hanover was an interesting mix of self-sufficient isolation and social and commercial interaction with the outside world. With the roads leading to Baltimore, Carlisle, York, Frederick, and Harrisburg (via Abbottstown), the town was a connecting link to several major markets. Just to the north and south of town, secondary roads led to Gettysburg and Westminster. Hanover was well situated to be a distribution center of the yield of the surrounding countryside. Taverns, inns, and hotels provided accommodations for teamsters and wagon traffic, while several blacksmith and carriage shops serviced the needs of the local citizens as well as those transporting goods to other markets. Cigar factories, carpentry shops, dry goods and grocery stores, along with numerous other businesses in the town provided work, as did the many productive farms in the area. (Almost all the Hanover Borough streets cited in this study still have the same names as they did in 1863. One exception is that the thoroughfare we know today as Broadway was generally

View from Hanover square to intersection of York and Abbottstown Streets. Although this view is commonly dated 1863, the Wentz, Overbaugh & Co. Dry Goods store opened for business in 1868. (GPL)

called Abbottstown Street, so I have used that historic designation. Concerning Frederick Street, this study employs two separate terms. For events within the town, I have utilized that same name. But for action that took place outside of the 1863 borough limits, I have usually referred to this route as the Hanover-Littlestown Road. Where other roads are concerned, I have generally used modern names for convenience.) [2]

The arrival of the railroad in 1852 expanded the local economic growth in several ways. The Hanover Branch provided efficient transportation and greater access to other markets. The depot and surrounding vicinity was transformed into an industrial and commercial center. Railroad agents, machinists and foundry workers, warehouse employees, and lumber and coal dealers turned much of the area into a veritable beehive of business activity. Telegraph lines along the railways now allowed up-to-the-minute communication with East Coast cities.

Even with all this economic activity, much of the area near the railroad depot was still undeveloped in 1863 including an open tract of land known as the Commons, which was situated between Carlisle Street and Abbottstown Street. This property had originally been twenty-three acres, as part of the estate of Col. Richard McAllister, and was sold by public subscription in 1798 to the people of Hanover. The Commons was a center of social activity; it was used at various times as a racecourse, picnic area, cow pasture, and space for local militia drills. Some of the land was developed by the railroad company for its station in 1852. In 1859, the Borough Council took possession of the remaining piece of open ground, having no idea then how vital a military role it would play a few years later.[3]

While the economic diversity of the town expanded, the German ancestry of many citizens continued to have a deep impact on the culture of Hanover. According to the census figures of 1860, more than 120 of the Borough's residents had actually been born in Germany; many more were first and second generation immigrants. The habits of the area, particularly farming and religious customs, strongly reflected this cultural heritage. By 1860, four churches were active in the town: the German Reformed, Saint Matthew's Lutheran, Methodist, and United Brethren. The names of a few surrounding townships such as Heidelberg and Manheim also pointed to a strong central European influence. In 1863, two English-speaking newspapers reflected the political leanings of the time: *The Spectator* was strongly Republican, while *The Citizen* was staunchly Democratic. (*The Citizen* also put out a German version. Another newspaper, *The Gazette*, was also published in German.)

In the months leading up to the war, great excitement had gripped many areas of the country. Once Southern states seceded from the Union, it did not take long for the emotions in many Northern towns to reach a fever pitch. Of immediate concern to many Hanoverians was the state of events in Baltimore. About forty miles to the south, this Maryland city contained large numbers of individuals friendly to the Confederate cause. When rioting broke out in that city on April 19, 1861, a state of alarm spread throughout Hanover. The greatest fear seemed to concern a gang in Baltimore called the "Rowdies," who were known to be Southern sympathizers. The common rumor was that the "Rowdies" were about to move on Hanover, rob the banks, and burn the town. A barricade was built across Baltimore Street consisting of wagons, boxes, lumber, and other implements. While some locals fled, many others armed themselves.

Although the "Rowdies" never invaded, the panic did provide some moments of local color. One man, Edward Steffy, was seen sitting on a front porch on Baltimore Street with a defective rifle. When someone asked what he planned to do with a useless gun, Steffy replied that "there was nothing wrong with the bayonet."[4] Other individuals served on picket duty, guarding and scouting on the roads south of town. Three local men were on guard along the Westminster Road one night when Isaac Wise climbed a tree as a lookout. When asked if he could see anything, his reply was a classic, "It's too dark, but it looks damn suspicious."[5] The "Rowdies" scare did not last long but later became a source of criticism and much amusement at the expense of many Hanoverians.

At the outbreak of the war, many local men began to volunteer for Federal service. Company G of the 16th Pennsylvania regiment was recruited locally. Before long, it became apparent that the war would last much longer than most had anticipated. After the three-month enlistment period of the 16th expired, many of those soldiers reenlisted in other units for longer terms of service. Other Hanover area men also began to volunteer in large numbers. The 76th, 87th, and 130th Pennsylvania had a number of locals on their rolls, while several other regiments contained at least a few Hanoverians.

Meanwhile, local citizens continued to follow the course of the war. Until late summer of 1862, Northerners could find some comfort in that the fighting had remained on Southern soil. But when Lee's Army of Northern Virginia crossed the Potomac in September of that year, a new state of alarm gripped many areas of Pennsylvania. Although no Confederate troops reached Hanover at that time, the campaign did have an impact on the local population. In August, Company C of the 130th

Pennsylvania had enlisted in Hanover. On September 17, 1862, these men went into battle near the village of Sharpsburg, Maryland, on what became the bloodiest day of the war. The 130th was part of the forces that attacked the sunken road that became immortalized as "Bloody Lane".

The first time that Gen. J. E. B. Stuart's Confederate cavalry made its presence felt on Pennsylvania soil was shortly after the Battle of Antietam. In October of 1862, Stuart conducted a cavalry raid in which he rode completely around the Union army. His force moved through Maryland and entered Pennsylvania west of the South Mountain. Before long, they crossed through the Cashtown Gap, turned south and rode through Adams County to make their way back to Maryland and into Virginia.

The years of 1861 and 1862 had provided moments of drama to various south central Pennsylvania communities. All that would pale in comparison to the events of 1863.

JUNE, 1863

The following editorial was printed in *The Hanover Citizen* newspaper, June 18, 1863:

> Much excitement has prevailed in our community for the last few days in regard to a raid into our borders of an immense force of Southern Cavalry. The severe skirmishing at a considerable distance from the main body of Lee's army leads us to believe that a forward movement on their part is designed. Yet such demonstrations may be intended as a feint to divert the attention of the Government away from the capital that an unsuspected army of large numbers under Lee may attack the Capital, are matters of doubt.
>
> One thing we would say to the people, that they shall be composed. It is not necessary that we raise an excitement that may, when the time comes to act, make us the degenerate sons of illustrious sires. It is not necessary that we prove ourselves cowards, by packing up our goods and running for our lives. Let us, as the protector of our wives and children, and our home in these our days be unmoved, and prepare to defend them. It is a singular way to show our devotion to the cause by leaving all to the mercy of the invader, without an effort individually to repel them. Suppose all were to take down their flags, move their goods, and flee to the hills, who would be left behind to tell that Hanover was a loyal town?
>
> Stand firm, ready to meet the worst, and if the whole thing proves to be a hoax, we can say at least that we were willing to do the best we could.

By late June of 1863, the fears expressed in this article had become reality as the main body of the Army of Northern Virginia had reached Pennsylvania. By June 27 of that year, the largest portion of Lee's force was located west of the South Mountain, and within a few miles of Chambersburg. Meanwhile, two infantry divisions had marched northeast toward Harrisburg, while one division, under Gen. Jubal Early, moved eastward through central Adams County. In his campaign report, Early stated that he was instructed to "proceed to York, and cut the Northern Central Railroad, running from Baltimore to Harrisburg, and also destroy the bridge across the Susquehanna at Wrightsville and Columbia..."[6] With Lee's army controlling a wide area of south central Pennsylvania, the physical and psychological impact of the invasion was then in full force.

By that time, the Army of Northern Virginia was moving through enemy territory without an effective cavalry screen. Some of their forces, however, were accompanied by small contingents of mounted forces. One of these units was the 35th Virginia Cavalry Battalion, under Lt. Col. Elijah White. This battalion, about 260 men, had been recently detached from Brig. Gen. William E. Jones's brigade; for several days, they had effectively been operating as a semi-independent command. By June 27, White had been ordered to report to Gen. Early. His Virginia cavalry would have a unique role in the expedition through Adams and York County.[7]

While Early's division continued its eastward trek toward York, the 35th Virginia Battalion was detached from this column to accomplish other important military objectives. White's mission on June 27 was specific. He was to make a raid on Hanover Junction and destroy railroad bridges and telegraph lines on the Northern Central Railroad. This destruction would prevent Union troops guarding Washington from reaching Pennsylvania by rail via Baltimore. The raid would also cut communications between Harrisburg and Washington. White's men were not on a foraging expedition but a quick strike. His troopers would have to move rapidly through Adams and York Counties and, after cutting these lines, rejoin Early's division the quickest way possible.

As the battalion rode eastward, they seized horses from various farms to replace animals that had broken down. John Rife and Daniel Geiselman were two of the several residents who had animals taken that day. Their farms were located west of the Conewago Creek (South Branch) on/near the road between Gettysburg and McSherrystown.[8]

Sometimes a soldier's lame horse was left in place of the healthy one. Substitutions such as this often occurred when a cavalryman had an immediate need for transportation. These "trades" differed from the full-scale confiscations, which both armies practiced. The 35th Virginia Battalion was a small body of troops; the unit did not have sufficient men to spare to control and move large numbers of extra animals. Also, their rapid movement did not allow a man to move very far away from the column to search for horses, lest he not be able to rejoin the unit.

The urgency of their mission, however, did not prevent these cavalrymen from confiscating other items along their route. Upon reaching McSherrystown, several soldiers crowded into the Reily and Sneeringer dry goods and grocery store. Sneeringer had not wanted to open the building but relented after the Confederates threatened to break in. In a damage claim filed to the Pennsylvania state government, the owners stated that the goods seized included the following: boots, shoes, hats, caps, undershirts, drawers, coats, jackets, sheets, cloths, hose, gloves, handkerchiefs, coffee, sugar, whiskey, soap, pen knives, steel pens, and stationery, for a total cost of $400. Neither owner mentioned whether the Southern men paid Confederate money for any of the items.[9] (See Appendix A for General Orders 72, General Lee's instruction for the confiscation of goods from Northern citizens.)

Around 10:00 A. M., the leading elements of White's battalion reached the northern outskirts of Hanover and moved southward on Carlisle Street. Daniel Trone, a telegraph operator for the railroad, was at his office at that time. The first sign of trouble was the sound of yelling and shooting. Trone looked out a warehouse window and saw several soldiers chasing two locals down a nearby alley. Another employee, William Stahl, watched as the Confederates pursued the two across the Commons, the open area near the railroad depot. Stahl claimed that the Virginians shot "at least a dozen times" as they raced after the men. One of the Hanoverians, Abdiel Gitt, escaped out the Abbottstown Pike. In the meantime, Trone had detached his telegraph instruments, carried them upstairs to an attic, and hid them under the floor.[10]

Before long, these Confederates returned. A few broke into the warehouse, but Gitt and Stahl had already left by another door. The two tried to make their way out of town by following the railroad tracks but found that a few troops maintained a guard along the Abbottstown Pike, which blocked that escape route. The men separated at this time; Gitt eventually made his way out of town by way of the Westminster Road. After walking about three miles, he obtained a horse and rode to

Westminster, then took a train to Baltimore.[11]

Days before, local railroad officials had heard of the approach of the Confederate army and made arrangements to move the locomotives and other cars to safety. One of the last, if not the only, pieces of rolling stock remaining in the Hanover freight yard was a handcar. Another railroad agent, Joseph Leib, and one of his co-workers made an escape by use of this manually powered device. As they moved along the tracks, two cavalrymen pursued, shot their rifles, and yelled for them to stop. Several hours later, Leib and his comrade reached Hanover Junction, and later moved to Baltimore.[12]

In Hanover, the main body of the 35th Virginia Battalion rode south on Carlisle Street and reached the square. The sight of the Southern soldiers caused fear among the townspeople, many of whom watched from their windows. The cavalrymen themselves were likely nervous also, wondering if any local militia were in the area. Confederate officers made sure no Union troops were in Hanover and sent guards out along the main street to the edge of town. Meanwhile, Lieutenant Colonel White addressed some of the residents in the square to assure them that no civilians would be harmed.[13]

One Southern officer rode up Baltimore Street to Peter Frank's blacksmith shop and requested that his horse be shod. The owner replied that he could not comply since it was a holiday. When the officer asked him what the occasion was, Frank brazenly replied, "the Johnnies were in town."[14] Frank's "holiday" ended quickly when the officer laid his hand on his holster. At this point Frank complied with cavalryman's request.

White's men stayed in the town for about an hour, while many of the soldiers entered local stores to obtain goods. How many of the items were paid for with Confederate money is unknown. Before long, the Virginians rode out the York Road and proceeded across York County toward Hanover Junction.

The arrival of the 35th Virginia Battalion had certainly provided moments of intense excitement. This was just a warm-up to what would happen a few days later.

CHAPTER ONE
STUART'S RIDE BEGINS

The embodiment of the Southern cavalier, James Ewell Brown ("Jeb") Stuart was one of the most colorful figures of the War of 1861-1865. After graduating from West Point in 1854, he served on the Kansas frontier for a time. Later, as an aide to (then) Col. Robert E. Lee, Stuart had been involved in the capture of John Brown during the Harper's Ferry raid. When Virginia seceded from the Union, he resigned his commission in the United States Army and entered Confederate service as colonel of the 1st Virginia Cavalry. Stuart quickly became noticed for his courage and was promoted to brigadier general on September 24, 1861. As his rank continued to rise, the numbers of cavalry under his command continued to grow. When he attained the rank of major general on July 25, 1862, he was placed in command of the Cavalry Division of the Army of Northern Virginia. Before long, his fame soared as he established a reputation as an outstanding intelligence-gathering officer. By 1863, Stuart had molded his force into a much feared and supremely confident fighting unit.[15]

Maj. Gen. JEB Stuart (NA)

In at least one regard, the typical Southern cavalry unit was at a disadvantage in comparison to its Northern counterpart. Union horsemen were generally equipped with a breech-loading single shot carbine, which, as the war went on, was often replaced with a repeating weapon. Confederate regiments often carried a mix of various shoulder arms. By mid-1863, some companies had acquired breech loading carbines, but others had to make do with Enfield rifles or various other types. A saber and revolver generally rounded out the cavalryman's armaments; Colt and Remington six-shooters saw widespread use.

Although the Confederate states could not match the industrial output of the North, in one area they had clear superiority. By the mid 1800s, a culture of horsemanship still predominated in many areas of the South. In the early stages of the war, the typical Confederate cavalryman was a far better rider than his Union counterpart. Throughout the first two years of the war, this excellence had been exhibited on many occasions, and Stuart's troops were the very personification of this superiority.

The Battle of Brandy Station, Virginia, was possibly the first major sign that the supremacy of Stuart's forces was to be tested. On June 9, 1863, Southern troops were surprised when Union horsemen splashed across Beverly Ford on the Rappahannock River in the early morning fog. After what turned out to be the largest cavalry action of the war, Northern forces eventually retreated back to the east side of the Rappahannock. However, Brandy Station was a shock to the Southern forces in more ways than one. Union troops had come perilously close to achieving a victory of significant size in the Confederate's "back yard." Even more galling was the fact that the despised Yankee cavalry had actually sought to battle with Stuart's men on a large scale. To add to Stuart's distress, the surprise nature of the attack caused comment in several Southern newspapers. Whether or not Stuart's contempt for his adversaries had been shaken, more evidence of a much-improved Union cavalry followed shortly. Several other mounted actions took place in the next few weeks as various elements of the opposing cavalry forces made contact in northern Virginia. Engagements at Aldie, Middleburg, and Upperville proved that the Union mettle and leadership shown at Brandy Station was no fluke. These fights had taken place as Lee's army marched northward while Stuart protected the Confederate infantry movements from the probing searches of the Northern cavalry. At this point, Stuart was still in communication with Lee, providing intelligence on the positions and movements of the Union forces. That situation was about to change dramatically.

On June 24, 1863, three brigades of Stuart's Cavalry Division gathered secretly under the cover of darkness near Salem Depot, Virginia. At about 1:00 A. M. on June 25, they began a movement which took them away from contact with their own infantry and into the rear of the Union Army of the Potomac.[16] Stuart's separation from Lee over the next several days effectively created a "communications blackout," which would shape the campaign in ways that no one in the Confederate command had imagined.

The idea for this movement seems to have originated from Maj. John Mosby. Mosby commanded a group of partisan irregular mounted troops that became famed for their operations in northern and central Virginia. In particular, their ability to conduct quick strikes on Union communication, supply, and railway lines was a constant embarrassment to the Federals. Mosby was also a potent intelligence gathering force within his sphere of operations. On June 24, he suggested to Stuart that instead of moving northward along the flank of Lee's main body, a way could be found through Virginia and Maryland by picking routes in between various Union infantry corps. Mosby believed the Federal corps were spread far enough apart that this movement could be accomplished without much difficulty.[17]

This idea was much to Stuart's liking. (In two previous expeditions he had actually moved completely around the Army of the Potomac; once during the Peninsula campaign, and the second time after the Battle of Antietam/Sharpsburg, Maryland.) The potential benefits of the mission were significant. By moving between various Northern forces, his men could capture supplies, and destroy railways and telegraph communications, while causing confusion as to the Confederate intentions.

After hearing this proposal, Lee decided to allow his cavalry commander the latitude to move northward however he saw fit. It is apparent, however, from dispatches sent by Lee that he believed it critical for Stuart to communicate with Gen. Richard Ewell's Infantry Corps promptly after he reached Northern soil. Lee's permission was given under the understanding that the cavalry would be able to rejoin the army before a major battle occurred in Pennsylvania.[18]

While this raid took place, two brigades of Stuart's Cavalry Division remained in Virginia for a time to guard specific gaps in the Blue Ridge Mountains. It was essential for the passes to be held by Confederate contingents. These troops would provide an early warning system in case Union forces made incursions toward the Blue Ridge. As long as these holding forces were in place, Confederate lines of supply and communication were safe as Lee's army marched northward behind the mountains.

Stuart's expedition began with about 5,000 men. The three brigades included were those commanded by Brigadier Generals Fitzhugh Lee and Wade Hampton, along with Col. John Chambliss, Jr. The commands under Lee and Hampton numbered about 2,100 and 2,000 troops, respectively. Chambliss's force, the smallest of the three, was around 1,300 strong.[19] In spite of recent casualties, these Confederate units were supremely confident, adequately armed veteran units.

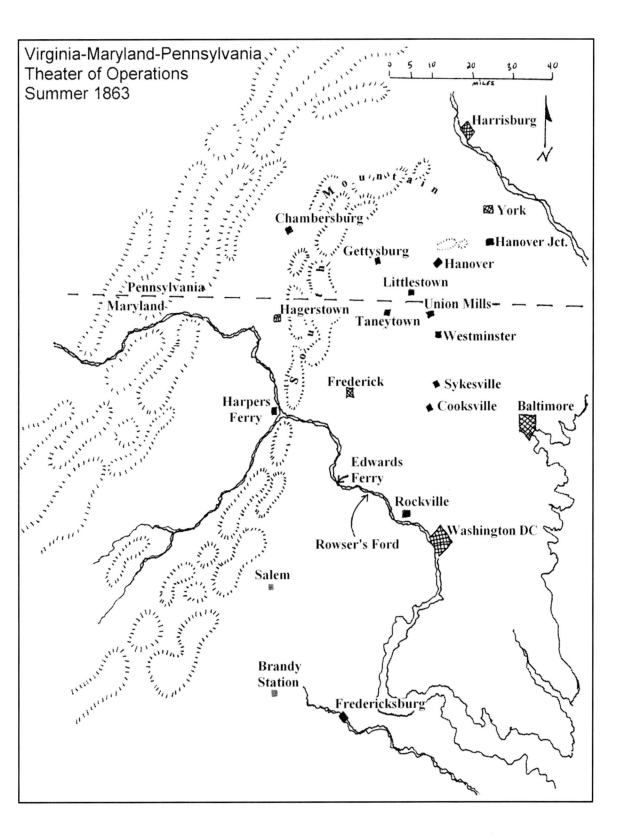

These brigades were led by officers in which Stuart had every reason to place great faith. Fitzhugh Lee, the nephew of Robert E. Lee, was born in 1835 in Fairfax County, Virginia. A graduate of West Point in 1856, he resigned his commission in May 1861 and entered Confederate service. In the early months of the war, he had served in various staff capacities. Before long, he was appointed lieutenant colonel, and later colonel, of the 1st Virginia Cavalry. As commander of that regiment, he was involved with the first of Stuart's expeditions around the Union Army, the "ride around McClellan" in the Peninsula campaign. Lee was then promoted to brigadier general on July 24, 1862.[20]

Wade Hampton, born in 1818, was a native of Charleston, South Carolina, and a graduate of South Carolina College. By the outbreak of the war, he was already a man of much importance in the South; he had served in his home state's legislature and owned immense land holdings. With his wealth and influence, he was able to organize the Hampton Legion, which he equipped at his own expense. Hampton was wounded at First Manassas as colonel of that unit; later he commanded an infantry brigade in the Peninsula campaign. On May 23, 1862, he was promoted to brigadier general; then in July of that year, he was assigned to command a cavalry brigade.[21]

FITZHUGH LEE   **WADE HAMPTON**   **JOHN CHAMBLISS**
(G & B)   (LOC)   (G & B)

John Chambliss, Jr. was born in Greensville County, Virginia, in 1833 and graduated from West Point in 1853. In 1861, he was engaged as a civilian planter, while also serving as an officer in a local militia unit. His first assignment in Confederate service was as colonel of the 41st Virginia Infantry. Then in July 1861, he became colonel of the 13th Virginia Cavalry.[22] In November of 1862, the 13th was attached to the brigade of Gen. W. H. F. "Rooney" Lee, Robert E. Lee's son. When "Rooney" was wounded at Brandy Station (and later captured), Chambliss was elevated to command the brigade.

Chambliss, Hampton, and Lee were experienced, aggressive officers with proven leadership qualities. It is clear that Stuart wanted his most capable commanders with him on the expedition.

The very first day of the movement Stuart discovered one big difference in comparison to his previous expeditions around the enemy: The Army of the Potomac was on the move. In fact, the Union Second Corps was actually marching along the very road that Stuart had intended to use. A few shots from the Southern artillery were not enough to persuade these infantrymen to take another route. Stuart was then faced with a choice that turned out to be one of the most critical in American military history — he could call off the movement and fall back toward the Shenandoah Valley. This option would forfeit the potential benefits of the raid but would ensure that the Confederate infantry had adequate cavalry forces to screen its movements into Pennsylvania. The other choice was to move much further east than originally intended, while still continuing the expedition. Stuart chose the second option. Instead of riding between the various Union corps, he decided to move completely around the Union army. His assessment was that he could complete this longer route and still rejoin Robert E. Lee's infantry before a major battle occurred. In his disdain for the enemy, Stuart was apparently guilty of the same mistake that Lee himself made, underestimating the initiative of the Union generals to move. It was a miscalculation they would both have much cause to regret.

The next day, these cavalrymen pushed through the villages of Fairfax Court House, Vienna, and Dranesville, already significantly delayed and much farther away from Southern infantry than Lee could have possibly imagined. By late night of June 27 and the early morning of June 28 the three brigades had reached Rowser's Ford on the Potomac River. With the swollen condition of the Potomac, the cavalrymen probably thought the term "ford" to be a cruel joke. After various attempts to locate an easy crossing spot failed, the crossing was made through a section about one half mile wide, with strong current and water that in some cases covered a rider's saddle. To transport the artillery ammunition across the river, the ammunition chests were emptied, and individual cavalrymen carried the artillery rounds on horseback to the Maryland side. One of Stuart's staff officers considered the crossing to be one of the more impressive feats that he had witnessed, stating that "No more difficult achievement was accomplished by the cavalry during the war."[23]

With the stress placed upon the animals by the strenuous river crossing, Stuart decided to rest the column for several hours before moving on.[24] After resuming the march, the cavalrymen reached Rockville, Maryland, where they promptly destroyed a telegraph line and gathered more supplies.

At this time, unaware of the Confederate presence, a Federal wagon train was approaching the Rockville area. Elements of Stuart's force were ordered to intercept the column. His official report provides a description of this incident, stating that "a long train of army wagons approached from the direction of Washington, apparently but slightly guarded. As soon as our presence was known to those in charge, they attempted to turn the wagons, and at full speed to escape, but the leading brigade (W. H. F. Lee's) was sent in pursuit. The farthest wagon was within only 3 or 4 miles of Washington City, the train being about 8 miles long. Not one escaped, though many were upset and broken, so as to require their being burned. More than one hundred and twenty-five best United States model wagons and splendid teams with gay caparisons [ornamental trappings] were secured and driven off. The mules and harness of the broken wagons were also secured."[25]

Every one of these wagons was loaded with provisions including oats, bread, hardtack, sugar, and hams. Capt. W. W. Blackford, the chief engineer on Stuart's staff, described how the contents of this train were of immediate benefit to the cavalrymen's horses. "... it did one's heart good to see the way the poor brutes got on the outside of those oats. After giving my horses all they could eat I slung half a bag, saddle-bag fashion, across my saddle for future use, and my horse seemed to know what this additional load was, for he occasionally turned an affectionate glance towards it."[26]

Now in control of more than 125 Federal wagons along with about 400 Union prisoners, Stuart once again was faced with a decision that would have critical consequences on the campaign. Various Southern officers would later state that the captured supplies should have been destroyed; it is not certain whether these concerns were raised at that time. Instead, it was decided to take the wagons along with the Confederate forces. Certainly the gathering of provisions did fall within the limits of the instructions given him by Lee. Apparently, Stuart believed that even with the logistical inconvenience of these extra wagons, he would still be able to move quickly enough to rejoin Confederate infantry and screen the right flank of Gen. Richard Ewell's Corps. Although it seemed a great prize at the time, the decision to keep the wagons would spark more controversy than any Stuart made during this campaign.

The capture of the wagons elicited a completely different response in Union command circles. The following dispatch reflects the anger and frustration of Quartermaster General Montgomery C. Meigs, in Washington D.C. The message was in response to a request by Brig. Gen. Rufus Ingalls for 10,000 pairs of shoes and socks needed by the Union army on the march:

General Ingalls,
    Chief Quartermaster, Hdqrs., Army of the Potomac:

    The bootees and socks will be ordered, and will be sent as soon as a safe route and escort can be found. Last fall I gave orders to prevent the sending of wagon trains from this place to Frederick without escort. The situation repeats itself, and gross carelessness and inattention to military rule has this morning cost us 150 wagons and 900 mules, captured by cavalry between this [place] and Rockville.

    Yesterday morning a detachment of over 400 cavalry moved from this place to join the army. This morning 150 wagons were sent without escort. Had the cavalry been delayed or the wagons hastened, they could have been protected and saved.

    All the cavalry of the Defenses of Washington was swept off by the army, and we are now insulted by burning wagons 3 miles outside of Tennallytown.

    Your communications are now in the hands of General Fitzhugh Lee's Brigade.
                      M. C. Meigs,
                      Quartermaster-General[27]

The expedition continued northward. During the night of June 28 and the morning of June 29 Lee's Brigade encountered a small Federal force near Cooksville, which retreated before them. Shortly after dawn, Lee reached Hood's Mill, where his men destroyed tracks of the Baltimore & Ohio Railroad, cut the telegraph line, and also wrecked a railroad bridge at Sykesville.[28]

In Stuart's official report of the campaign, he stated that at this time, "The enemy was ascertained to be moving through Frederick City northward, and it was important for me to reach our column with as little delay as possible, to acquaint the commanding general with the nature of the enemy's movements, as well as to place with his column my cavalry force."[29] This information on the Union troops clearly would have added an extra sense of urgency to Stuart's movements.

By late afternoon on June 29, the leading elements of the column reached Westminster. Unknown to Stuart, about ninety-five Union men of the 1st Delaware Cavalry Regiment had been ordered there to guard the railroad station. Westminster was also a critical road junction in the event of an attack on Baltimore.

As the Confederates approached the town, they were greeted by a surprise attack when Company C of the 1st Delaware charged southward on Main Street toward the vanguard of Stuart's column. The Southerners recovered quickly, and the 4th Virginia began to push Company C back through the town. In an attempt to stem the tide, Company D of the 1st Delaware was thrown into the fray. Before long, the Virginians began to surround and overwhelm the small band of Union troopers. In danger of being encircled, the Delaware men attempted to escape southward along the Reisterstown Road, pursued closely by Confederates. By the evening of June 29, Companies C and D of the 1st Delaware Cavalry had essentially ceased to exist. The official report of the unit listed sixty-seven men missing, which included a small detail from the 150th New York Infantry on duty in the town as a provost guard.[30]

The events at Westminster were as critical as any on Stuart's ride. The skirmish and chase, along with the movements earlier that day, required the cavalrymen to stop and care for their horses. According to Stuart, this was "the first time since leaving Rector's Cross-Roads" that his men "obtained a full supply of forage."[31] The night of June 29 and the morning of June 30 found Stuart's men encamped from Westminster northwards along the Baltimore-Littlestown Pike, with Lee's Brigade in advance near the Union Mills area.[32]

Even with all the delays that had taken place, Stuart believed the original premise of the movement was still attainable. He hoped to reestablish communications after reaching Pennsylvania the following day. That night, however, another cause for concern came to his attention when a scout reported the presence of Union cavalry at Littlestown, Pennsylvania, about six miles to the northwest of Union Mills. Once again enemy troops had been discovered in a place Stuart had not anticipated.[33] The stage had been set for the battle at Hanover, Pennsylvania the next day, but neither opposing commander was aware of the drama about to unfold.

CHAPTER TWO
UNION

The Battle of Chancellorsville also caused profound changes in the Union Army of the Potomac, but for not very positive reasons. Once again, Lee had beaten another Union general on Virginia soil. In the heated atmosphere and debate that followed this defeat, several individuals emerged as scapegoats. According to many key individuals in Washington and most of the Union corps commanders, army commander Gen. Joseph Hooker was considered the main culprit of this debacle. But from the viewpoint of Hooker himself, one of the individuals deserving the most censure was Gen. George Stoneman, commander of the Army of the Potomac's Cavalry Corps. When Stoneman went on sick leave, Hooker promoted Gen. Alfred Pleasonton to the command of his cavalry and made no secret of the fact that he did not want Stoneman back.

Pleasonton, no stranger to political intrigue himself, immediately began to pull strings to add additional troops to his cavalry force. He particularly had his eye on the division of Gen. Julius Stahel, which had been operating independently from the Army of the Potomac as part of the defenses of Washington. Pleasonton wanted this unit to be placed under his command and, before long, convinced Hooker of their necessity. On June 28, 1863, Stahel was relieved of his command and his troops were assigned to the Army of the Potomac. That same day, this division was placed under the leadership of Gen. Hugh Judson Kilpatrick.[34]

Meanwhile, a much bigger controversy had arisen over the status of Hooker himself. Whether he knew it or not, his time to command the Army of the Potomac was quickly coming to a close. After Chancellorsville, doubts and outright hostilities concerning his leadership had rapidly escalated, and several powerful individuals were looking for any possible reason to have him dismissed. Hooker's most powerful antagonist was his superior in Washington, General in Chief Henry Halleck. The two generals were so opposed that earlier that year, a special "arrangement" had been in place whereby Hooker reported to and received his orders from President Lincoln himself, without having to deal with Halleck. By summer, Lincoln apparently realized that this bypass in the usual chain of command was having some negative effects. On June 16, Lincoln sent a dispatch to Hooker in which the president asserted that Hooker was under the direct command of Halleck.

A few other Union generals had been informally approached regarding the possibility of taking army command, but each had declined. Lincoln was now one of the few supporters Hooker still had in Washington, although even he was beginning to have strong concerns regarding Hooker's ability to command a group of officers so openly hostile to his leadership. The president found himself in the awkward position of realizing that most of the Union generals opposed Hooker, yet no one wanted to command the army himself.

As the Confederates marched northward, Hooker moved the Army of the Potomac in response, keeping an inside track between the Confederates and Washington. Most Union infantry crossed the Potomac River at Edwards's Ferry on June 26, and by June 28, much of Hooker's force was positioned around the vicinity of Frederick, Maryland.

Within the framework of this campaign, certain issues between Hooker and Halleck began to escalate tensions to an explosive level. After months of mutual distrust and anger, the subject that finally brought matters to the point of no return was the debate over the Federal garrison at Harper's Ferry.

Hooker believed that to coordinate an effective plan against Lee, it was essential for him to have control of not only the Army of the Potomac, but all Union troops within his sphere of operations.

Also believing himself to be outnumbered by Lee, he asked to have the Union troops at Harper's Ferry (approximately 10,000 men) released to his command. When this request was denied, Hooker sent a message to Washington stating that under the command restraints which he had been placed, he could no longer perform the duties required of him. Whatever the general's expectations were regarding this message, the response of the administration was to immediately accept his resignation. Hooker's leaving set the stage for one of the most critical decisions faced by any United States president in a time of war. In the midst of the crisis, President Lincoln turned to Gen. George Meade.

Maj. Gen. George Meade
(LOC)

Unlike other Union generals, Meade was not given a choice on the subject of army leadership. In the early morning of June 28 a courier from Washington appeared at his Fifth Corps headquarters and handed him the order that placed him in command of the Army of the Potomac. Rarely in world history has an individual been placed in such intense pressure so suddenly as George Gordon Meade. By late June, the invasion was in full force; the Army of Northern Virginia was well into Pennsylvania. Yet in the midst of a campaign that many believed might decide the nation's fate, the Lincoln administration had placed a different officer in charge of its largest army. Meade's role, like Hooker's, was a dual one. He had to aggressively go after Lee's forces while still keeping Union troops in a position to defend Washington. By this time, much of the Army of the Potomac was concentrated around Frederick, Maryland. On June 28, Meade began to make the strategic and operational decisions to move his army into Pennsylvania, continuing the pursuit of Lee.

Meade assumed command knowing that a battle was likely imminent, and accurate information regarding Lee's movements was essential. His use of Pleasonton's Cavalry Corps would be a large factor in the campaign. As the army marched northward through central Maryland, the mounted forces formed a moving protective arc that preceded the infantry. The cavalry's role was also dual in nature: obtain vital intelligence while also screening the Union infantry marches. Gen. John Buford's First Division guarded the army's left and front as they probed along several roads, mostly east of South Mountain. The army's right and rear were protected by the Second Division under Gen. David McMurtrie Gregg. The Third Division, now under command of Gen. Hugh Judson Kilpatrick, covered the area between the two previously mentioned units, as they screened the army's center and right.

Hugh Judson Kilpatrick typified the aggressiveness of Union officers promoted in the reorganization of the Union cavalry. After graduating with the West Point class of 1861, he had served as a captain in the 5th New York Infantry. He was the first Unites States Regular Army officer to be wounded in the war, at Big Bethel, Virginia. In September of 1861, he was promoted to lieutenant colonel of the 2nd New York Cavalry, and in December of 1862 attained the rank of colonel of that regiment. On June 14, 1863, Kilpatrick was promoted to brigadier general of volunteers. While working his way up through the ranks, he had taken part in numerous Eastern theater actions.[35] Now, with the Confederates moving northward, Kilpatrick commanded a cavalry division that had a critical role in the campaign.

Within Kilpatrick's division, a few other notable promotions took place. On June 29, 1863, the two men who would command brigades in his command jumped from junior officer status directly to general, bypassing several intermediate ranks. Brig. Gen. Elon J. Farnsworth assumed command of the First Brigade, Third Division, Cavalry Corps. Although only 25 years old by this time, Farnsworth had served as an officer in the 8th Illinois Cavalry and spent some time on the staff of General Pleasonton.[36]

The Second Brigade in Kilpatrick's Division was given to the charge of Brig. Gen. George Armstrong Custer who, at 23, was even younger than Farnsworth when he was made a general officer. (This promotion was even bigger, because Custer was technically a first lieutenant at that point. He had temporarily served as a Captain prior to that time.) Custer had not particularly excelled at West Point, graduating last in the class of 1861. He had come close to being expelled more than once in his four years at the academy and was actually under detention when he graduated. But as undistinguished as his academic record was, his military career was just the opposite. During the first two years of the war, he had served on the staffs of several general officers including George McClellan, Alfred Pleasonton, and Philip Kearny and had already been noted for conspicuous bravery several times.[37]

Although politics played a big part in their rise to general officer status, Custer and Farnsworth were brave and daring leaders. Their promotions were a prime example of the overall transformation that was taking place within the Federal forces at this time. As various organizational and personnel changes occurred during the campaigns of 1863, officers of greater competence were taking positions of increasing responsibility.

Hugh Judson Kilpatrick **Elon J. Farnsworth** **George A. Custer**
(LOC) (G & B) (NA)

Meanwhile, several less aggressive commanders were being placed in capacities farther away from the dominant theaters of the war. The Confederates were drawing closer to battle with an enemy unlike the one they had faced previously: a Union force that was being led much more skillfully, at several different levels, than the army they had faced on numerous Virginia battlefields.

Kilpatrick's official report stated that when he assumed command of the Third Division, its "actual strength" was 3,500, "although it numbered on paper upward of 4,000 men for duty."[38] After General Stahel's transfer, the division had been restructured from three brigades down to two.

Farnsworth's First Brigade included the 1st (West) Virginia, 1st Vermont, 5th New York, and 18th Pennsylvania regiments. (The 1st West Virginia was often referred to as the 1st Virginia since West Virginia had just recently attained statehood.) Although, in name, these units were from four different states, the brigade was even more "mixed" than it appears. The 1st West Virginia not only had troops enlisted from that state but also had significant numbers from the southwestern counties of Pennsylvania and the southeastern region of Ohio.

The 18th was recruited in the late summer and early fall of 1862 and was the newest of the four regiments. The other three units enlisted in the early stages of the war, and had served in various areas where they saw action in numerous skirmishes. More recently, these troops had spent time in northern Virginia as part of the defense forces of Washington, D. C., and had sustained a few casualties in small affairs with Mosby's Rangers. On paper, these regiments were each composed of twelve companies. A few companies, however, were on detached duty and were not present with the brigade that morning.[39] The regimental numbers ranged from the 1st West Virginia, with slightly more than 400, to the 1st Vermont, with possibly 600. This likely made the overall battle-ready strength of Farnsworth's brigade that day close to 2,000 men.[40]

The majority of the brigade was equipped with a Sharps carbine as their primary weapon. The exception was the 18th Pennsylvania, and portions of the 1st West Virginia, that had Burnside carbines. A typical sidearm for most of these men was a Colt .44 caliber revolver, although some of the 1st Vermont carried the Remington .44.[41]

Accompanying this brigade was Battery E, 4th United States Artillery, commanded by Lt. Samuel Elder. This unit consisted of four three-inch rifled guns, and about sixty-five men on duty.[42]

While Farnsworth's First Brigade was composed of three veteran regiments, along with one that was untested, Custer's Second Brigade was almost the reverse. Custer's command consisted of the 1st, 5th, 6th, and 7th Michigan Cavalry regiments. The 1st was the most battle hardened of the four. It had been organized very early in the war, and had acquired an outstanding reputation for its fighting in several engagements. The 5th, 6th, and 7th, however, had not taken part in any major actions up to that time, although they had taken part in scouting expeditions in Virginia, and some of their companies had been involved in various skirmishes. Since these last three regiments had not been in the service for long, the numbers of this brigade were somewhat larger than most. The battle strength of Custer's brigade on June 30 was probably about 2,300, with the 5th Michigan alone having more than 700 men.[43]

While the pistols carried by these Michigan men were generally Colt .44 revolvers, their shoulder arms were a rather interesting mix. Many of these troopers carried either a Sharps or Burnside carbine. But the 5th Michigan, and at least four companies of the 6th, had been outfitted with Spencer rifles.[44] The Spencer was more accurate than the shorter barreled carbines used by the great majority of cavalrymen. But its real advantage was the rate of fire. With metallic rim-fire cartridges loaded into a seven-shot tubular magazine, the Spencer had a distinct advantage.

Like Farnsworth, Custer also had an artillery unit traveling with his men at that time. Just assigned to the Michigan Brigade was Battery M, 2nd U. S. Artillery, commanded by Lt. Alexander Pennington, Jr. Pennington's battery was composed of six three-inch rifled pieces, and around 120 men present for duty.[45]

Before Custer's promotion, this brigade crossed the Potomac River on June 25 at Edward's Ferry and had been performing intelligence gathering duties under Brig. Gen. Joseph Copeland. In fact, the 5th and 6th Michigan had already reached Gettysburg by June 28, where they discovered that Early's Confederate division had marched through that town two days before. These cavalrymen then rode back into Maryland and rejoined other Union troops near Emmittsburg. Meanwhile, the 1st and 7th Michigan had scouted in the direction of Sharpsburg, Maryland, and Harper's Ferry.[46]

Before Kilpatrick could utilize his new division to its peak efficiency, he first needed to establish contact with all the units that had been transferred to his command. However, it would still be a few days before all his troops would be gathered into a unified whole. With the Army of the Potomac moving northward, Kilpatrick had to develop his division's role as part of a moving cavalry screen before he had even met some of his regimental commanders.

On June 29, the majority of the division moved northward from the vicinity of Frederick, Maryland. Their destination that day was Littlestown, Pennsylvania. Farnsworth traveled the most direct route through Woodsborough and Taneytown, Maryland; his men reached their objective after sunset. Custer, with the 1st and 7th Michigan, rode through Utica and Creagerstown to Emmittsburg, then over to rejoin Farnsworth. According to Dexter Macomber of the 1st Michigan, his regiment reached Littlestown at 11:00 P. M. to "stop for the night."[47] Meanwhile, the 5th and 6th Michigan were still several miles away. These two regiments had fallen back into Maryland after their scouting foray to Gettysburg. They continued to patrol through the countryside on June 29th, and did not reach Littlestown until the early morning of June 30th.[48]

During this campaign, Union soldiers found a welcome reception in Pennsylvania towns. According to Chaplain Louis Boudrye (5th New York), when the troops reached Littlestown "we were received with the greatest demonstrations of joy by the people. A large group of children, on the balcony of a hotel, waving handkerchiefs and flags, greeted us with patriotic songs.... How different was such reception from that we had been accustomed to have given us by the inhabitants of Virginia villages!"[49] Charles Blinn of the 1st Vermont was extremely impressed with the women of the town. Blinn's diary entry for June 29 includes the following: "About fifty young ladies had assembled at the Union Hotel and as we entered the place, sweet singing greeted our ears and made glad our hearts. The Red, White, and Blue Star Spangled Banner was sung with great enthusiasm, the chorus being joined in by hundreds of weary soldiers. May the fair ladies and good people of Littlestown long live in peace, prosperity, and happiness."[50]

At some point, Kilpatrick received his marching orders for the following day. Since the Twelfth Infantry Corps had been ordered to reach Littlestown on June 30th, the Third Cavalry Division was to move well in front of them, in the direction of Hanover and Abbottstown. A few days before, information had reached Union headquarters that Early's Confederate division had moved eastward through Gettysburg. Southern infantry marching from Gettysburg toward the Susquehanna River would likely pass through Abbottstown, so a movement by Kilpatrick toward that town would fulfill two major responsibilities. His cavalry would likely gain information on the Confederates, while he continued to screen Union infantry.

Upon Kilpatrick's promotion, Companies A and C of the 1st Ohio Cavalry were assigned to be his headquarters guard.[51] This squadron had previously performed the same service for General Stahel, their former division commander. During the night of June 29, a small detail of that unit was sent to scout in the direction of Abbottstown. This patrol, under Lt. John McElwain, reached the vicinity of (East) Berlin and captured a few Confederate stragglers of Early's division.[52] Kilpatrick was gaining more information on what was in front of him. The next day he would find out, to his surprise, what was behind him.

CHAPTER THREE
EARLY MORNING JUNE 30

The homes of Andrew and William Shriver of Carroll County, Maryland sat on opposite sides of the road from Westminster to Littlestown; the sympathies of the families were on opposite sides of the war. The Southern loyalties were deep enough in the William Shriver household that six of his sons served in the Confederate forces before the war ended. Ironically, Union sympathizer Andrew owned a few slaves, while William did not. The night of June 29 was the start of at least a week of intense excitement for both households as a steady procession of troop movements through their land would create disruption on a huge scale. Over the next several days, the emotions of these families ranged from joy to concern and fear, depending on which troops were moving through the area of Union Mills, Maryland at that time.

After the engagement at Westminster, the leading forces of Stuart's troops began to reach the Shriver property along Pipe Creek, shortly after 10:00 P. M.[53] Before long, Fitzhugh Lee had fallen asleep under an apple tree in the Andrew Shriver orchard. As the main body of his brigade gained some much needed rest near Union Mills, the leading elements had advanced north of there, sending scouts in the direction of Littlestown and Hanover.

After a few hours of sleep, Stuart arrived in the area near Pipe Creek by "about daylight" on the morning of June 30th.[54] The general's presence heightened the excitement considerably in William Shriver's home as the place became filled with Confederate officers eating breakfast. Afterwards, Stuart joined in a spirited singing session around the family piano. The festivities were long remembered by those present, but before long, the Confederate officers returned to the business of the day. The major concern at this point was to reestablish contact with the Confederate infantry. Stuart had captured enough supplies and destroyed enough railroad and telegraph lines to consider that portion of his movement a success.

Whatever Stuart's original intentions were, after he learned of the Union cavalry near Littlestown, he decided to move in the direction of Hanover. By heading north toward Hanover, he hoped to pass to the east of the enemy and communicate with Gen. Richard Ewell's infantry corps. Stuart believed that reaching a gap in the Pigeon Hills north of Hanover was a key to this movement.[55] What he did not know was that on the morning of June 30, Kilpatrick's cavalry division was also headed toward that same town.

A study of maps of Carroll County, Maryland, and York and Adams Counties, Pennsylvania, reveals a triangle with Union Mills at the southern tip, Hanover at the north, and Littlestown at the westernmost point. It is along the perimeter and, in some cases, through the interior of this "triangle of operations" that the opposing forces moved toward their unintended meeting at Hanover.

The distances to be covered and the nature of the roads themselves were critical factors. Since Kilpatrick's troops had bivouacked east of Littlestown, their movement to Hanover was likely about six miles and covered relatively flat terrain. Meanwhile, Stuart's main body negotiated a longer route that morning. Although Lee's men had slept near Union Mills, the other two brigades had encamped south of there. As a result, the main column march with Chambliss and Hampton was close to twelve miles. Worse yet, much of this movement would take place on the Westminster-Hanover Road, a virtual roller coaster route of several steep hills. Lee's Brigade, meanwhile, had the task of screening the column. They traveled on other roads to the left of Chambliss and Hampton, between the main body and Littlestown.

Chambliss's Brigade led the Confederate main column. After them came the captured wagon train, then Hampton's Brigade. A typical cavalry movement covered approximately four miles per hour. But the hills along the route proved difficult for the wagons. This terrain caused a gap between the front and rear units which had a profound impact on the engagement that day and Stuart's decision making. Chambliss had the smallest of the three brigades and was headed toward an encounter in which any quick reinforcement by other Southern forces would be impossible.

MAIN COLUMN MOVEMENT

Unlike the Union cavalry, the Confederates maintained the pre-war structure of ten companies per regiment. Chambliss's Brigade, composed of the 2nd North Carolina, along with the 9th, 10th, and 13th Virginia, was organized in this manner. Not all of these regiments, however, had their full complement at this time. According to one officer, four companies of the 9th had been detached a few days before, one of which did not rejoin the regiment "until the fall."[56] The overall brigade strength that morning was around 1,300.[57]

Early in the morning of June 30, Chambliss's men rode northward past the Shriver Mill, then began their trek along the Westminster-Hanover Road. A few Confederate accounts describe the movement beginning at "dawn" or "daybreak."[58] The 13th Virginia was the leading regiment in the line of march and was followed by the 9th Virginia, 2nd North Carolina, and 10th Virginia.[59] Also accompanying the column were a few artillery pieces. At least two cannon, and probably four, moved with Chambliss. (See chapter six and Appendix E for more on the Confederate artillery.)

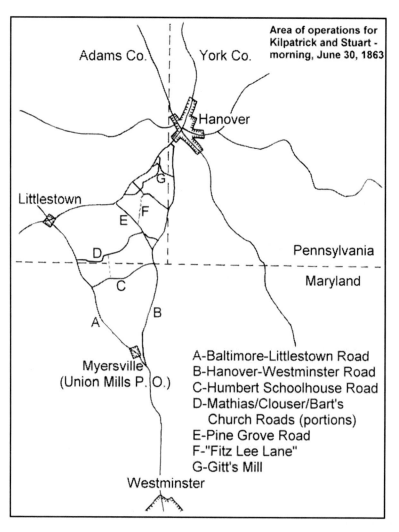

Herbert Shriver, one of William's sons, rode with the Confederates that day. Apparently, Stuart and Fitz Lee had taken quite a liking to the boy. In exchange for his services as a guide, Stuart agreed to secure admission for Herbert to the Virginia Military Institute.[60]

Soon after turning onto the Hanover Road, the column began to climb a long incline, just the first of many hills they encountered along that particular road. The Maryland portion of this route was particularly strenuous, with a number of grades of varying length and steepness.

About four miles later, the column reached the Pennsylvania line and "details were immediately sent out to seize horses."[61] Adam Leese was one of the first Pennsylvanians to suffer losses from Stuart's main column. Two mares and one horse were taken off the Leese family farm, near the intersection of the Hanover-Westminster Road and the State line. Before reaching York County, the Hanover-Westminster Road traverses a two-mile stretch of the southeastern corner of Adams County. At least seven other Adams County residents had animals or goods seized just within this two-mile stretch.[62] The taking of horses was just beginning to get into high gear.

The following excerpt likely reflected the feelings of many Confederates at this time. One soldier wrote:

> The march was resumed at dawn next morning. An order detailing a squad of men and an officer from each regiment to collect horses for our dismounted men satisfied us that we had passed from Maryland, and had entered the State of William Penn, whose armed sons we had so often seen upon the soil of our native Virginia. The time had come to pay back in some measure the misdeeds of men who, with sword and fire, had made our homesteads heaps of ruin, and, in many instances, left our wives and children not a horse, nor cow, nor sheep, nor hog, nor living fowl of any kind. Soon a country store was reached and trooper after trooper escaping from the ranks quickly filled it with Confederates, who, without asking the price, were proceeding to help themselves to any and every article they needed or fancied. The first field officer, however, who discovered what was going on, rode quietly up and cleared the store, compelling the men to put back what they had taken, and posted a guard to remain until the command had passed.[63]

This account likely refers to the business owned by John Baublitz, whose general store sat along the Westminster Road in Union Township, Adams County. Baublitz later filed a damage claim amounting to $243 for losses that included dry goods, groceries, and cash.[64]

Soon after passing the Baublitz store, the column reached the Adams/York County line. The first York County farm that Chambliss's men encountered that morning was the one owned by Josiah Gitt. Gitt operated a dry goods store on the Hanover square and owned a few other parcels of land near the town. In 1863, Edmund and Catherine Lippy were working his farm along the Westminster Road. By the time the main Confederate column had passed by, Gitt had lost two mares, one stallion, three mules, seventy-five bushels of corn, and twenty bushels of oats. (Property was also seized by Federal troops the next day. On July 1, the Union Fifth Corps marched along the same road, and another mare was taken along with a wagon saddle and other farm gear.)[65]

As the main body moved toward Hanover, the search for horses in the surrounding countryside continued. According to one Southern officer, each regiment sent out one detachment to seize horses, with each detachment consisting of one officer and five men. Capt. William A. Graham (Company K, 2nd North Carolina) was in charge of one of these details. Graham's men did not rejoin their regiment until the main action was over, but they certainly had a great impact on several locals.[66] These details were particularly active along Grand Valley Road and Impounding Dam Road. (postwar names)[67] The gathering of horses was of vital concern for Stuart's cavalry. Since their movements had already consumed much more time and mileage than expected, the wear and tear on horses and men was becoming severe and would shortly become unbearable, as soldiers and animals were strained to the breaking point. Unfortunately for Stuart, each delay and detour added to the hardships

Modern view of the Josiah Gitt farm. Gitt lived in Hanover but owned property outside the town. This farm was worked by a tenant, Edmund Lippy, at the time of the battle. Electrical poles to the right of the house are situated along a portion of the Westminster Road which still follows its original course. However, in 1863 the road ran directly past the house and barn and continued to the left of this view. The construction of Long Arm Dam caused a portion of the road to be diverted around the lake. (This West Manheim Township farm is sometimes confused with the Gitt's Mill site. The mill was actually owned by Jeremiah Gitt and was located in Union Township, Adams County approximately one mile northwest of here.) (AC)

and created even more demand for procuring healthy mounts. It is not certain how many local citizens received Confederate cash or receipts for their animals. Since Stuart's troops were on the move, the formalities of payment were often ignored, especially by detachments well away from the main force. In some cases, horses were seized from fields and stables without the owner being present. Certainly in these instances, the soldiers did not have the time to search for the owner, even if they had been inclined to do so.

The following letter, although written by a Union cavalry officer during another campaign, gives an indication of the hardships suffered by the cavalry on both sides:

> ...you have no idea of their sufferings [the horses]. An officer of
> cavalry needs to be more horse-doctor than soldier, and no one who
> has not tried it can realize the discouragement to Company commanders
> in these long and continuous marches. You are a slave to your horses,
> you work like a dog yourself, and you exact the most extreme care from
> your Sergeants, and you see diseases creeping on you day by day and
> your horses breaking down under your eyes, and you have two resources,
> one to send them to the reserve camps at the rear and so strip yourself of
> your command, and the other to force them on until they drop and then
> run for luck that you will be able to steal horses to remount your men,

and keep up the strength of your command. The last course is the one I adopt. I do my best for my horses and am sorry for them; but all war is cruel and it is my business to bring every man I can into the presence of the enemy, and so make war short. So I have but one rule, a horse must go until he can't be spurred any further, and then the rider must get another horse as soon as he can seize on one.

To estimate the wear and tear on horseflesh you must bear in mind that, in the service of his country, a cavalry horse when loaded carries an average of 225 lbs. on his back. His saddle, when packed without a rider in it, weighs not less than fifty pounds. The horse, in active campaign, is saddled on an average about fifteen hours out of the twenty-four. His feed is nominally ten pounds of grain a day and, in reality, he averages about eight pounds. He has no hay and only such other feed as he can pick up during halts. The usual water he drinks is brook water, so muddy by the passage of the column as to be the color of chocolate. Of course, sore backs are our greatest trouble. Backs soon get feverish under the saddle and the first day's march swells them; after that, day by day the trouble grows. No care can stop it. Every night after a march, no matter how late it may be, or tired or hungry I am, if permission is given to unsaddle, I examine all the horses backs myself and see that everything is done for them that can be done, and yet with every care the marching of the last four weeks disabled ten of my horses, and put ten more on the high road to disability, and this out of sixty-[equivalent to] one horse in three.

Imagine a horse with his withers swollen to three times the natural size, and with a volcanic running sore pouring matter down each side, and you have a case with which every cavalry officer is daily called upon to deal, and you imagine a horse which has still to be ridden until he lays down in sheer suffering under the saddle. Then we swipe the first horse we come to and put the dismounted man on his back. The air of Virginia is literally burdened today with the stench of dead horses, federal and confederate. You pass them on every road and find them in every field, while from their carrions you can follow the march of every army that moves.[68]

Not only were animals being impressed but, in some cases, people as well. Each unforeseen detour and delay caused another dilemma as Stuart was being forced away from primary routes to use lesser-known, secondary roads. His forces had great need of local guides, and some York County men were forced to ride along with the movement; their specific knowledge of the roads was a valuable asset to the Southern horsemen. Ephraim Nace operated a mill in West Manheim Township and said that after Southern cavalrymen seized his four horses, "One of the rebels said that General Stuart wanted me to go along too, and asked me to mount my best horse and ride with them. I did not want to go, but they forced me to accompany them." Nace also mentioned that "several other persons" were taken along to act as guides including Jacob Leppo and Samuel Keller.[69]

Like several other Pennsylvania residents, Jacob Leppo had been persuaded by unscrupulous individuals to take the oath of the Knights of the Golden Circle, a shadowy organization with Confederate sympathies. The Knights worked to exploit anti-Union sentiment and attempted to incite opposition to the war among Northern citizens. Some locals were told that if they flashed a specific secret sign, the Confederates would not take their property. Whether this organization actually had a viable presence in this area is debatable, but some opportunists preyed on the fears of local farmers and charged them a fee for this false hope. Later in the day, Leppo talked to Stuart himself. The general told him that this sign "was simply a money making business on the part of the members of the society" and had no validity.[70]

As the horse gathering details roamed over several back roads of southwestern York County, Chambliss's men were approaching Hanover. Meanwhile, the wagon train and Hampton's Brigade were further back, negotiating the demanding terrain. After crossing the South Branch of the Conewago Creek (sometimes locally known as the Little Conewago), the main column encountered the southern face of Conewago Hill. This steep slope was one of the last significant inclines they had to deal with that morning. Shortly after, the forward elements of Chambliss's Brigade reached some of the high ground southwest of Hanover, from where they were able to see the town. Also in sight, and more unexpected, was a small Union detail moving across the Confederate front, on another road to Hanover. Stuart's ride had brought the vanguard of his column upon the rear of Farnsworth's brigade.

KILPATRICK LEAVES LITTLESTOWN

Early that morning, the units of Kilpatrick's division also began to ride toward Hanover. The 1st and 7th Michigan departed the Littlestown area well before any other regiments; their march probably began before first light.[71] Also traveling with this portion of the column were the six three-inch rifles of Pennington's Battery M, 2nd U. S. Artillery. General Kilpatrick and his escort were also on the road early. (Although most historians have held that he rode at the front of the 1st Michigan, he may have been at the head of Farnsworth's brigade, which moved out somewhat later.)[72]

About an hour after the 1st and 7th Michigan left Littlestown, Farnsworth's Brigade began to move, led by the 1st West Virginia and 1st Vermont. (See Appendix D for a discussion on the order of these two units.) Following these two regiments were the four 3-inch rifles of Elder's Battery E, 4th U.S. Artillery. Behind them, the 5th New York and 18th Pennsylvania completed the line of march of Farnsworth's brigade.[73]

Meanwhile, the other regiments of Custer's brigade did not travel with the main body at that time. The 5th and 6th Michigan used Littlestown as a base of operations for a few hours, as some of their men performed scouting missions to the south of that town.[74]

In contrast to the demanding terrain encountered by Stuart's main column, Kilpatrick's men experienced a relatively easy movement to Hanover that morning. Their route was a fairly level one, with a few rolling hills but no steep grades. About four miles from Littlestown, the Union troopers reached the South branch of the Conewago Creek and the mill owned by John Duttera. At that time, the building was often called Kitzmiller's Mill, in reference to its original owner. The original structure had been built in 1738 and was said to have been the oldest mill in Pennsylvania west of the Susquehanna River.[75]

Not long after crossing the Conewago, Kilpatrick's column reached York County. The Union men encountered several residents as they rode along, and individual acts of kindness by some of the locals left quite an impression on the soldiers. George Hoch, of the 18th Pennsylvania, never forgot the consideration shown by Henry Sell, whose farm sat along the Hanover-Littlestown Road near Plum Creek. Hoch recalled that Sell "kindly cared" for the men and horses as they passed his land. Even years later, members of that regiment were saddened to hear of Henry's death when they came back to visit the area.[76]

Not far from the Sell property, Samuel and John Forney were working their own family's land when the troops rode by. The two brothers not only conversed with a few Union men but also with one captured Confederate.[77]

At least one other young boy got a close look at the Union column near that area. George Spangler had been leading a few cows to pasture, but after spotting the approaching cavalry, he decided to let the cows take care of themselves while he sat on a fence and watched the action. A few of the advanced guard had passed before an officer rode up and spoke to him. The Union man greeted him kindly and asked if George could accompany him into Hanover. The lad stated, "I would rather remain on the fence because I want to see all the soldiers pass by and I might miss them if I go into town." The officer convinced him by saying that since all the soldiers would pass through the town anyway, he could just as easily see them at that location. As they moved along, the officer asked him if he knew where Mr. Jacob Wirt lived; the boy replied that he would point out the house when they arrived. The two continued to talk as they proceeded into town. Shortly afterwards, George found out that he had escorted General Kilpatrick into Hanover.[78]

The sight of Union troops was an immense relief to the townspeople after being visited by Confederates a few days before. Citizens began to emerge from their houses to see the spectacle and cheer the cavalrymen. The weather likely contributed to the occasion; at least two locals described the morning as a "beautiful" day filled with sunshine.[79]

With the guidance of George Spangler, General Kilpatrick reached the Wirt residence on Frederick Street. (Jacob Wirt was not home at that time. As president of a local bank, he had taken many of the bank's holdings away from Hanover for safety.) Once inside, the general began to evaluate the next stage of the movement. Of particular interest to him was a map of York County hanging on the wall. Calvin Wirt gave the map to Kilpatrick to help guide his troops.[80]

Shortly after, some prominent citizens entered the house including Reverend William Zieber of the German Reformed Church. When Zieber heard Kilpatrick mention the tired and thirsty condition of the troops, he walked out to the street and encouraged the residents to provide food and water for the soldiers. Within minutes, local citizens began to bring forth an assortment of food and drink for the cavalrymen.[81]

The 1st and 7th Michigan did not remain in the town for long. In short order, these two regiments continued northward toward Abbottstown. Meanwhile the head of Farnsworth's brigade reached Hanover, around 8:00 A. M.[82] Farnsworth's troops halted in the streets of town where they were able to enjoy the local hospitality. By this time, the streets were overflowing with civilians, and the soldiers continued to receive food and gifts. A joyous atmosphere prevailed, and the event quickly turned into the most unforgettable party the town had ever seen. According to one soldier of the 1st West Virginia, "Flags and handkerchiefs were waved, patriotic songs were sung, and it seemed a time of general rejoicing."[83] Charles Blinn (1st Vermont), who had been so impressed with the women of Littlestown, seemed to have romance on his mind at Hanover also. He wrote in his diary that "The misses with bouquets of sweet flowers were making impressions on the hearts of good looking soldiers and every thing went merry as a marriage ball."[84] Most soldiers were probably more focused on the food. Horace Ide (1st Vermont) said that "The good people brought out food of all kinds, such as pie, cake, coffee, and so forth, and we were enjoying ourselves immensely..."[85] According to another Vermont man, "Matrons and maidens and children ran with bread and milk, beer and pretzels. Dinners hot from the fires were brought to the tables before the doors, and, to a hungry homesick army, it was a scene perhaps unsurpassed in the marches of the war."[86]

By about 10:00 A. M., most of Kilpatrick's men had already reached or passed through the borough limits of Hanover. But with the 1st and 7th Michigan near Abbottstown and some of the 5th and 6th still scouting south of Littlestown, there was a distance of at least twelve miles between the front and rear of Kilpatrick's division.[87] (Kilpatrick himself had also left Hanover by that time and continued toward Abbottstown.)

Between these two "sections" of Michigan troops, Farnsworth's brigade "straddled" Hanover at that time. The head of the brigade was one or two miles north of the square (likely near the village later known as New Baltimore).[88] From there the 1st West Virginia and 1st Vermont stretched along the Abbottstown Road, but at least some of the Vermont men were still in the northern section of Hanover. (See Appendix D.)

The square was still very crowded, with many soldiers of the 5th New York around that area and along Abbottstown Street. Evidence seems to indicate that Elder's Battery was also in this general vicinity.[89]

Meanwhile, the 18th Pennsylvania had paused along Frederick Street, with most of the unit in the southwestern edge of town. Also traveling with this portion of the column were a number of wagons including at least a few ambulances.[90] Accounts suggest that while the main body of the regiment was in front of these wagons, at least some cavalrymen were immediately behind them.[91] Meanwhile, further out on the road toward Littlestown, another small contingent of about forty Pennsylvanians acted as a rear guard for the column. The companies of the 18th were dispersed over a much greater area than the other regiments. That deployment created a situation in which even the most battle tested unit would have a difficult time responding to any surprise attack. When their rear guard was spotted by the leading elements of Chambliss's Brigade, the 18th Pennsylvania was in the worst possible spot at the worst possible time.

CHAPTER FOUR
ENCOUNTER

As Kilpatrick's division moved toward Hanover, various small details guarded the flanks and the rear of the main column. Unfortunately, the actions of most of these patrols have been lost to history with very little documentation written by any of the individuals involved. The accounts of Henry Potter, a Second Lieutenant of Company M, 18th Pennsylvania, provide a notable exception. About forty men of Companies L and M of the 18th were placed under Potter's command to act as a rear guard for the marching column. His men had been ordered to remain "about a mile in the rear" as the column progressed that morning.[92] While the main body of the 18th Pennsylvania relaxed in Hanover, Potter's detachment approached the small collection of houses around the intersection of the Westminster and Hanover-Littlestown Roads.

At this time, the leading elements of Stuart's forces approached along the Westminster Road. Two conflicting versions relate what happened at this time. According to Potter, a detail of "about fifty" Confederates rode out in front of the rear guard and demanded that his detachment surrender. Instead, the lieutenant ordered his men to move slowly toward the Southerners, and on his signal, they fired and charged through the Virginians. Although this strike initially scattered the Confederates, they recovered quickly and raced after the fleeing Pennsylvanians. (Writing well after the war, Potter believed the detachment he charged was from the 13th Virginia.)[93]

Col. Richard Beale of the 9th Virginia gave a different perspective of the same incident. Beale stated that the 13th Virginia led the Confederate march, followed by two squadrons of the 9th. The Southern cavalrymen spotted "a squadron of the enemy" that "advanced slowly up the road in our front. The major commanding the Thirteenth Regiment, seeming to hesitate, Lieutenant Pollard [of the 9th Virginia] was ordered to the front with his squadron to charge the enemy. This was gallantly done, and the Federals, breaking, ran back into Hanover, followed by our whole force."[94]

No matter which side took the initiative in this encounter (perhaps a mutual attack was made), the Union rear guard dashed up the road toward Hanover with Virginians in pursuit. Soon after, Potter's detachment came into contact with other 18th Pennsylvania men who were dismounted along Frederick Street, in the southwestern edge of town. These soldiers, after hearing the firing behind them, had begun to scramble to mount their horses and prepare their rifles. At this point, Potter shouted to his men to "Right about" to face the enemy. His rear guard, along with other supporting troops, then charged back toward the Confederates and began to push the 9th Virginia squadron (and possibly, elements of the 13th Virginia) back the Littlestown Road. Potter believed that about one hundred more Pennsylvanians joined in this initial counterattack.[95]

While the rear companies of the 18th Pennsylvania charged back in the direction of Littlestown, at least one officer from another Union regiment was also in the action. Lt. Alexander Gall, an adjutant of the 5th New York, was apparently detached from his regiment when the first shots were fired. He raced up to Potter and shouted questions about what had happened. Within the next few seconds, a Confederate bullet struck down Gall as he charged along with the Pennsylvanians.[96]

For a short time, the momentum shifted as the small contingent of Virginians fell back. This running gunfight continued along the Hanover-Littlestown Road and extended a short distance onto the Westminster Road. The resistance of the Confederate detail began to stiffen as they retreated toward their own supporting troops. The combat was mostly at close quarters at this point; various fences and hedgerows restricted much of the fighting to the roads themselves. According to Lieutenant Potter, this stage of the battle involved "cutting and slashing," and "with those of us who

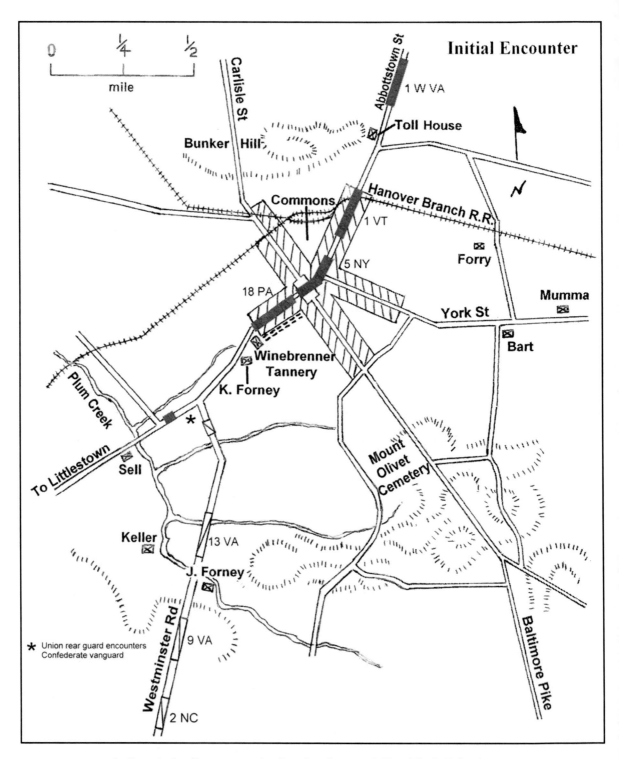

1- Potter's detail encounters leading detachment of Chambliss's Brigade near intersection of Westminster and Littlestown Roads.
2- Small unit action flows back and forth on Hanover-Littlestown Road.

were in the front it was a fist fight."[97] Somewhere near the intersection of these two key roads, this movement came to a standstill as the Pennsylvanians' momentum stalled.

Potter's rear guard and a few other companies had made it through this initial encounter relatively intact, but things were about to get much worse. As the flow of the fight reversed its direction on the Hanover-Littlestown Road, the Pennsylvanians were moving closer to the presence of a much larger Confederate force. Worse yet, the farther they rode from Hanover, the greater the separation became between themselves and the rest of their own regiment. During the march that morning, the deployment of the regiment had placed some of its companies in a vulnerable position. Now, because of their own aggressive response to the initial encounter, they were in a much more perilous situation.

Only a few minutes had elapsed since the first shots were fired. At about that time, the real problems for the 18th Pennsylvania began as Southern troops launched the first major attack. Charging northward along the Westminster Road, the 13th Virginia struck the Union men near the Littlestown Road intersection.[98] The 13th, possibly supported by elements of the 9th Virginia, quickly gained the upper hand.

By this time, a few Confederate artillery pieces had reached the field and opened fire in support of the assault. These guns were initially positioned on high ground near the Westminster Road. Although their numbers were small, the cannon may have had an effect well beyond actual casualties. The 18th Pennsylvania had never before been in a major battle. This was not only the first time they had come up against any frontline Confederate cavalry, but it was also the first time they had been exposed to artillery fire. Considering their inexperience, the psychological impact of the Confederate guns was likely more important than the physical.

Under intense pressure, a good part of the 18th began to dissolve. Some Pennsylvanians began to flee in the direction of McSherrystown, while others raced toward Hanover with Virginians in hot pursuit. According to Lieutenant Potter, about that time "our men gave way... I lost my horse then, and was using my carbine...." He further added that as his men were overrun, "I laid on the ground close to the fence and played dead" until all the charging Southern men had passed.[99]

As this running gunfight moved once again toward Hanover, more chaos started to break out in the town itself. To avoid capture, some ambulance wagon drivers began to rush their teams along Frederick Street. These wagons started a chain reaction of confusion. The panic-stricken horses plunged into other troops of the 18th Pennsylvania and disrupted much of their attempt to form an organized resistance.

By that time, the rear companies of the 18th had already disintegrated under the pressure, and the Pennsylvanians were about to have another problem. Another Confederate regiment was about to join the battle. Lt. Col. William H. Payne, the commanding officer of the 2nd North Carolina, described what happened about this time. Payne stated that while the initial attack had "succeeded in cutting in twain the Eighteenth Pennsylvania Cavalry, which brought up the rear of the Federal forces under Kilpatrick, it did not succeed in driving the Union forces through the streets of Hanover and out of the town. It was then that General Stuart, mounted on his horse Virginia, rode up to me and asked if I would lead a charge into the town

Lt. Col. William H. F. Payne
(G & B)

and drive the Yankees out."[100]

The 2nd North Carolina likely crossed Plum Creek while still on the Westminster Road. At some point, they left that road or the Hanover-Littlestown Road to approach the southwestern edge of town. To launch their attack, the unit needed an unobstructed route that would enable them to strike the flank of the Union column in Hanover. Apparently, a dirt lane/alley that ran parallel with Frederick Street served this purpose. Payne was ordered by Stuart to "take the outside street, penetrate deep into the body of the town & cut off the force." The regiment had around 150 mounted men in this assault. As the North Carolinians rode closer to the enemy, Payne remembered the following scene: "As each cross street gave us a glimpse into...town, we saw it packed with masses of cavalry... I suppose we had passed half the length of the town before my advance turned to the right and attacked, and the attack was... so vigorous as to throw the enemy near us into utter confusion."[101] (Payne probably meant to say that his men turned left, unless he was referring to turning right onto Frederick Street itself. See Appendix F for more on the attack route and regimental strength of the 2nd North Carolina.)

Much of the 18th Pennsylvania still had their hands full as they tried to hold off the increasing pressure of the 13th Virginia (and possibly elements of the 9th Virginia). Now this additional strike by the 2nd North Carolina shattered another portion of the Pennsylvanians' column. Panic ensued as the Southern horsemen struck their flank. The sheer force and suddenness of the attacks pushed fleeing soldiers into other companies of the regiment which spread further confusion. As the shock of the assault pushed closer to the center of town, some of the 18th Pennsylvania began a chaotic retreat up Frederick Street. Adding to the mayhem, some wagon drivers continued to rush their teams headlong up the street, which made it almost impossible for company commanders to mount an effective resistance.

Not all of the teamsters retreated without a fight. Individuals from various regiments were assigned as ambulance guards, and some fought to protect the wagons. One small detail of the 5th New York, under Sgt. Bradley Alexander, immediately made a counterchage when the Confederates swarmed into the column. As Alexander led his men into a mass of the enemy, he received a saber blow to his head, which knocked him off his horse. In the midst of the pandemonium, one of the wagons ran over him, inflicting severe injuries to his back and kidneys. Several other men from Companies E and F, 5th New York, became casualties as they fought to save the ambulance vehicles.[102]

The initial fighting at Hanover provides a classic example of how an individual soldier's knowledge of battle is often limited to his own small, intense area of combat. The accounts of the 18th Pennsylvania cavalrymen reflected a wide range of experiences depending on the location of their individual companies. Capt. J. B. Smith of Company B stated that his men "were driven through the town in a jam" and made no mention of being involved in any counterattack.[103]

Meanwhile, Sgt. George Hoch, Company E, said that after retreating to the Market House, at least some of the Pennsylvanians halted and reformed and were involved in driving the Confederates out of town.[104] Another Pennsylvanian had an entirely different experience. In the flank attack by the 2nd North Carolina, Lt. Thomas Shields and a detail of about twenty-five men were overrun. Most were likely captured quickly, but he became separated from his unit in the melee and would find that day that his adventures were not quite finished.

Stuart's official report did not state his precise location when the fighting started. The account of his chief ordnance officer, Capt. John Esten Cooke, suggests that the first shots were fired before the general reached the field.[105] Stuart most likely would have preferred to allow the Union column to pass without a fight; his reason for moving to Hanover was to avoid enemy cavalry. But the initial

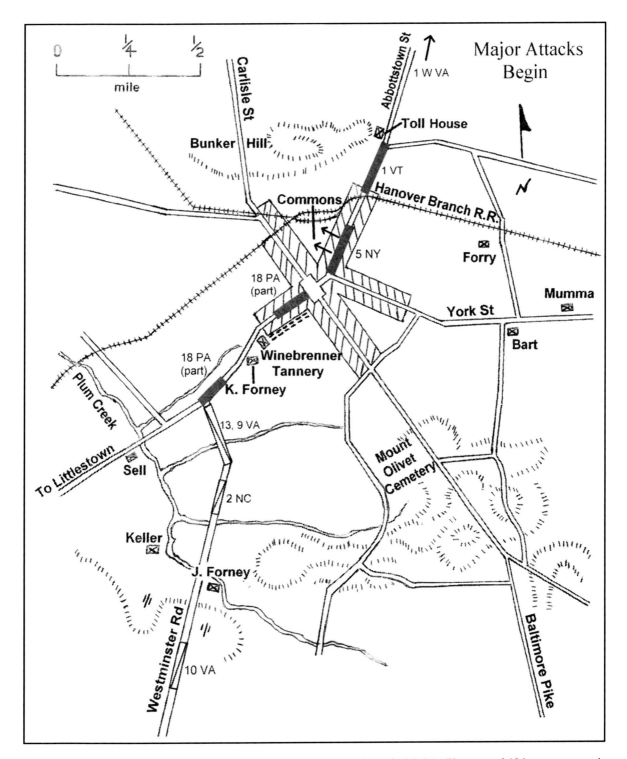

1- Major Confederate assaults begin as 13th VA smashes rear portion of 18th PA. Elements of 18th retreat toward McSherrystown, others toward Hanover with Virginians in pursuit. **2-** Confederate artillery (likely one section of McGregor's battery) opens fire from high ground along Westminster Road. Their initial positions may have been closer to Frederick St. **3-** 2nd NC advances, then strikes the flank of front companies of 18th PA by charging along "alley" and side streets. **4-** 5th NY moves to Commons area and forms for counterattack.

encounter, along with aggressive action taken by Colonel Chambliss, brought on the battle. Stuart's official report includes the following: "About 10 a.m. the head of the column reached Hanover, and found a large column of cavalry passing through, going toward the gap of the mountains [Pigeon Hills] which I intended using. The enemy soon discovered our approach, and made a demonstration toward attacking us, which was promptly met by a gallant charge by Chambliss' leading regiment..."[106] There was no rebuke by the general concerning the action taken by Chambliss; apparently the initiative shown by the colonel was deemed appropriate, possibly even unavoidable, under the circumstances.

The Union presence in Hanover that morning was a critical blow to Stuart's plans. His forces could have moved beyond the Pigeon Hills by taking either of the roads to Abbottstown, Carlisle, or York, but only if he first controlled the road junction in Hanover itself. Adding to this dilemma was the secondary road system in the area. At that time, there was no practical way for large numbers of troops to bypass the town without significant delays.

Once the fighting started, Stuart likely determined that the best course of action was to strike quickly and rout the enemy. But he had no idea of the number of Union soldiers in the vicinity of Hanover. When the battle broke out, the 18th Pennsylvania was probably the only Federal regiment within his sight and was clearly vulnerable. After the initial assault by Virginia troops, some of the 18th was already disintegrating. Stuart possibly sensed a window of opportunity. If the 2nd North Carolina could rout the remaining Union forces in Hanover, possibly this would clear the route that Stuart had intended to use.

When the fighting reached the town, many locals were stunned at the turn of events. As the realization of what was taking place began to sink in, the atmosphere of joy quickly turned to fear. One 5th New York veteran described it this way: "The sudden surprise had a strange effect upon the people, who at first did not realize what had happened. Then came the sounding of bugles, the hurried orders, the quick movements of the troops, and the fierce rebel yell which sounded louder and louder as the enemy came charging on and on into the town. Hearing and seeing all this they began to realize that this was war, cruel war. The women and children fled in terror from the scene, seeking shelter within their homes. What a change! In less time than it takes to tell it, the streets were full of live Rebels who penetrated to the very centre of town, even to the market place."[107] (This marketplace was a pavilion-like structure in the center square and was mistaken for a covered bridge by some soldiers who charged under its roof.)

Rebecca Scheurer described herself as "one of the children in the throng" in the Hanover square that day. When the shooting started, the young girl was "pulled from between artillery horses by my mother, and hastening home, I secreted myself in a rear room, and realized something of the shock of war, as I heard the whizzing of bullets over the house."[108] Some cavalrymen assisted in helping local children to safety. S. J. Mason of the 5th New York lifted several children over a fence and encouraged them to run home to their basements.[109]

One five-year-old boy, Ambrose Schmidt, was with his sister in a small backyard when the fighting started. It is not certain what the two children actually saw of the battle, but apparently they became frantic with fear, and their imaginations ran wild. Ambrose claimed that they ran back into their house "crying that we saw two men cut off each other's heads."[110]

The ultimate objective in a mounted assault, even more so than causing casualties, was to destroy the cohesion of enemy formations. The most effective way to accomplish that goal was to maintain discipline and momentum in the attacking force. Up to that point, Southern troops had performed that mission admirably. A few minutes after the 2nd North Carolina charged, significant numbers of Confederates had reached the center of town, and elements of the regiment had made it a few blocks north of the square. A large portion of the 18th Pennsylvania had been shattered, and most of its companies had ceased to be effective. At this point, an observer might have concluded that things looked ominous for the Union cavalry. A bigger view, however, would have revealed a different scene. Union troops were already reacting to the emergency.

Maj. John Hammond
(BBGB)

One Union soldier whose courageous actions were especially notable that day was Maj. John Hammond, commanding the 5th New York regiment. When the shooting started to the southwest of town, the major rode along Frederick Street and urged the citizens to move inside for their own safety. He also quickly began to take tactical control of the situation by turning much of his regiment to the left and directing it along a side street toward the Hanover Commons. By moving off the main street before the force of the Confederate attack reached his regiment, Hammond was able to keep some unit cohesion, even as he began to prepare his men for a counter attack. Because of his clear headed thinking, instead of the Confederate momentum slamming into the 5th New York in the area of the square, the Southerners soon found themselves on the receiving end of the blow. Although Stuart did not know it at that moment, his hopes of controlling the road through Hanover were about to come to a sudden end.

The 5th New York was an experienced, well disciplined unit. The men kept their composure and moved into the open area of the Hanover Commons, where they quickly prepared for battle. It was not long before they launched their counterattack. According to one soldier, Hammond ordered the column "down a side street toward the railroad depot, and formed into line upon a vacant lot. Breaking by fours, he ordered and led a charge with drawn sabres. We met them at the market-house, and were instantly engaged in a fierce hand-to-hand conflict."[111]

The counterstrike of the 5th New York in the center of town was sudden and dramatic. The fighting was not only fierce but in many cases close and personal, with sabers, pistols, and fists being used freely. After a few minutes of this incredibly violent melee, the force of the New York onslaught became too much for the Southern men. As Union pressure increased, the unit cohesion that Confederates had maintained in their own charge began to fall apart. Before long, greater numbers of Virginians and North Carolinians began to flee along Frederick Street back to their own supports. Many Southern cavalrymen continued to fight as they fell back within various "pockets of resistance." Several alleys, streets, and backyards were turned into personal theatres of combat as individuals and groups fought for their lives in their own small scale battles. Fighting also spilled over into various crop fields, as some Southern horsemen tried to reach the cover of the hills south of town, where a few of their own artillery pieces were still engaged.

Much of the 5th New York charged past the offices of Hanover's two English speaking newspapers. Employees of *The Spectator* witnessed some of the action and were impressed with the fighting of several New Yorkers. *The Spectator* noted that Capt. [Charles] Farley was observed as he "gallantly" led a portion of the charge upon the enemy. Newspaper workers also saw men of the 5th capture what they described as a "Rebel squad, 15 in number," apparently quite close to their office.[112]

The 5th New York played, by far, the major role in pushing the Confederates out of the center of town. Meanwhile, other troops provided assistance. George Hoch, Company E, 18th Pennsylvania,

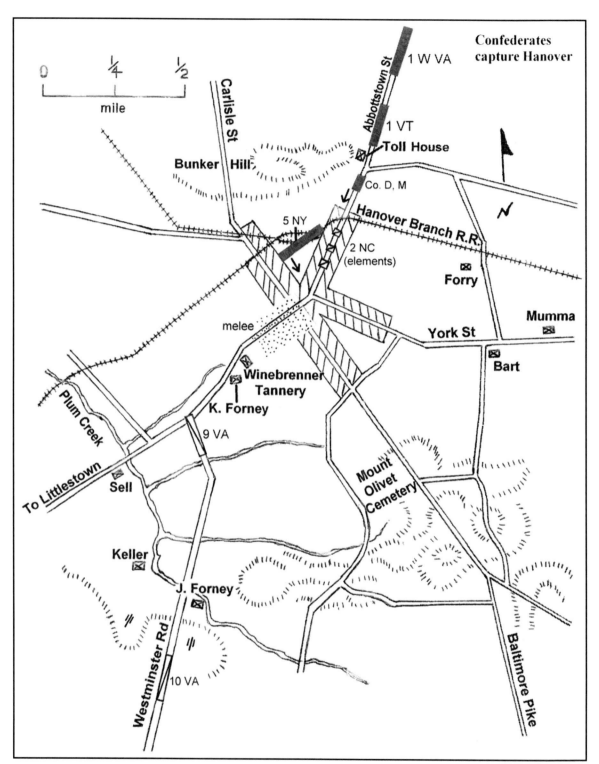

1- Much of the 18th PA has been shattered. Some companies of the regt. continue to fight on Frederick St. and side alleys.
2- Confederates control the center of town. Advance elements of the 2nd NC reach the railroad crossing on Abbottstown St. **3-** 5th NY deploys in the Commons area. They charge from there and strike Confederates in and about the square. Companies D and M of the 1st VT charge advanced elements of the 2nd NC along Abbottstown St.

later recalled the death of one of his comrades. Hoch said that John Hoffacker "was a few yards in advance of the Company, and was admonished by Sergeant Jacob Greenswalt not to be too fast, to which Mr. Hoffacker replied: 'Oh, God, I am too close to my home to die.' Almost instantly a pistol ball pierced his forehead, killing him instantly. He was a brave man, of good habits, and had many friends."[113] (See Appendix H for more on John Hoffacker.)

The area near the Market House was not the only place where Union troops struck back. While the 5th New York had been deploying in the Commons in preparation for their strike, leading elements of the 2nd North Carolina had already progressed a few blocks past the center of town. Rev. William K. Zieber watched some of the fighting and stated that Southern forces "drove our troops out Abbottstown Street to the railroad."[114] The area of the Abbottstown Street (Broadway) railroad crossing could be called the Confederate "high tide" of the Battle of Hanover. The forward elements of the 2nd North Carolina that charged past the square soon began to lose momentum as they approached the railroad tracks. It is likely that some of the Southern men were already falling back toward the center of town before any Union counterattacks reached this vicinity. Several cavalrymen were trying to bring off Union prisoners or captured wagons, but very shortly these North Carolinians were fighting to escape capture themselves.

By this time, the Confederates had lost much cohesion and momentum, and the men leading the charge were about to run into a bigger problem. Although they had charged past the center of town, other Union troops were still in their front, farther out on Abbottstown Street.

The 1st West Virginia and 1st Vermont were well north of the square when the fighting started. (Their precise locations at that time are difficult to establish; see Appendix D.) One Vermont man stated that "the head of the column was nearly through the town" when the first shots were fired.[115]

Modern view of Broadway looking north toward railroad crossing. Elements of the 2nd North Carolina reached this area before Union counterattacks. (AC)

Civil War era view of Abbottstown street looking south, German Reformed Church on right. (GPL)

Another 1st Vt. soldier, Horace Ide, remembered that "we were enjoying ourselves immensely when suddenly some distance behind us, bang went [a] cannon. Some one says 'the folks are firing a salute in honor of our arrival.' Just then the shell burst and we came to the conclusion that people didn't generally fire shells for a salute..."[116]

Responding quickly, Ide's Company mounted their horses and "the column moved on at a walk."[117] At least some of the 1st Vermont and 1st West Virginia continued to ride northward along the road for a short time. Company commanders could then establish some command and control while they awaited further orders. Some troops maintained column formation until they reached a more open area, where they were then able to maneuver more efficiently and deploy before counter charging. As these Vermont soldiers moved along Abbottstown Street, several fleeing Pennsylvanians actually rushed by them.

Upon hearing the sounds of battle, Union officers reacted decisively. General Farnsworth himself ordered the 1st Vermont and 1st West Virginia back toward Hanover. He specifically

48

directed a squadron (two companies) of the 1st Vermont to help repulse the attacks. Horace Ide claimed that when his company was near the toll gate, they became aware of the presence of Confederates behind them; the order was then given to companies D and M to "Right about wheel" in preparation to charge.[118] (This toll gate is shown on the 1860 York County map and the 1876 Hanover area map on the west side of Abbottstown Street [Broadway] very close to the intersection with the road now known as Ridge Ave. Ide's statement confirms that at least some companies of the 1st Vermont were still within and/or very near the town limits when the first shots were fired.)

As the companies of the 1st West Virginia and 1st Vermont deployed, at least one officer used the presence of the troops on Northern ground as a motivation. According to one 1st West Virginia man, Maj. Charles Capehart shouted, "Remember boys that we are on the free soil of Old Pennsylvania, with Stars and Stripes to cheer us on to battle. We will drive the Rebels off her soil."[119]

Any thoughts the North Carolinians had about holding the area north of the square could not have lasted long once they saw Union troops racing in from the north. As the counterattacks increased in number, some Confederates were in danger of being completely cut off from their own supports. When the 5th New York emerged from the Commons and struck Confederates in the square, the leading elements of the 2nd North Carolina not only had Union men in their front, but also behind them. Lt. Col. Payne believed that "had there been no Yankees behind us we might have maintained ourselves until relieved, or escaped.... however, they poured out from every street. My command was cut in two."[120] The North Carolinians were then forced to escape from the town any way possible.

By that time, the Southern assault had gained much ground but lost significant force. The men in the front of their attacking column were especially vulnerable; the leading companies of the 2nd North Carolina had lost much of their original density. According to Horace Ide, when the Vermont men counterattacked along Abbottstown Street, "the Rebel column had come some distance and were pretty well scattered, so that they turned and fled back through the town and we after them."[121]

As companies D and M of the 1st Vermont made their way back through the town, the 5th New York (and elements of the 18th Pennsylvania) were still fighting the 2nd North Carolina and 13th Virginia south and west of the square. Although the center of Hanover was once again under Union control, much of Frederick Street was still a very unstable area. Because of the fluid nature of the battle, unit cohesion was lost as the fighting degenerated into a whirlwind of hand to hand melees occurring in the streets, alleys, and surrounding fields. The experience of Lt. Col. Payne was typical of many soldiers in this phase of the fighting. After the Union counterattack, Payne described being "engaged for some minutes in a confused melee, in which I remember nothing but firing pistols, clouds of dust, an occasional thump with a saber; a sudden collapse of my horse tumbling dead headlong under me; jumping to my feet and being knocked down by a Yankee's horse, crawling out on my hands and knees as the crowd swept by. One of my men ran ahead of me into a tan house and I, under some vague hope that a countercharge would recover us, followed him."[122] Although he was out of the battle, Payne's role that day was not yet completed.

The two Vermont companies added their strength to the 5th New York and helped to regain control of the town and some of the open fields between Baltimore and Frederick Streets. The Vermont men captured about twenty Confederates, but it is not certain whether these Southern soldiers were taken in the initial counter strike on Abbottstown Street or in fighting south of the square. Meanwhile, the remaining companies of the 1st Vermont arrived shortly after and deployed, for a short time, in a supporting position near the center of town.

The 1st West Virginia had also been ordered back toward the fighting, but by the time they arrived, almost all of Hanover was once again in Union hands. This regiment maneuvered through the area south of the square and took position on the left of their brigade. Most of the unit moved into the open ground just west of Baltimore Street, where they deployed in line of battle.[123]

While these units deployed, Union artillery reacted to the situation as well. Battery E, 4th U. S. Artillery, had traveled with Farnsworth's brigade that morning. This unit, commanded by Lt. Samuel Elder, was somewhat unique among the Federal artillery. Most Union batteries had six cannon; Elder's had only four.[124] When the battle started, these four three-inch rifled guns were in, or very near, Hanover. A rise of ground, which later became known as Bunker Hill, was the obvious choice for their deployment.[125] This elevation runs roughly east to west and bisects the Carlisle Road about half a mile north of the center of town. The area provided room to maneuver, great fields of fire, and was one of the few prominent positions that offered relative safety from Confederate attacks. After reaching the hill, Elder's guns began to target Confederate troops south of the town. They remained there for most, if not all, of the fight and were later supplemented by Pennington's battery.

Although the 5th New York had regained control of Hanover, its forward momentum was stopped momentarily. To the southwest of town, the New Yorkers encountered elements of Chambliss's Brigade that had not been involved in the initial Confederate charges. Major Hammond reported that after driving the Confederates out of town, "we found a large force drawn up in the road as a reserve and received from them a severe fire, causing the men to halt for a moment."[126] This Confederate "reserve" was a large portion of the 9th Virginia and possibly regrouped elements of the 13th Virginia. Another Union man used the same adjective as Hammond in his description of the intensity of this fighting. Horace Ide (1st Vermont) stated that "our men followed the North Carolinians out of town, but there the rest of the column was met and quite a severe fight ensued."[127]

One individual who was conspicuous for his bravery at this point was Sgt. Selden Wales, Co. A, 5th New York. Wales had been at the head of his company as the 5th New York charged on Frederick Street. When the Union counter offensive stalled, he attempted to rally his men to move forward. At this point, Wales fell from his horse; he had been shot through the heart.[128] The 5th New York had no shortage of heroes that day.

If the battle's outcome still hung in the balance, it would not be so for long. About that time, General Farnsworth reached Hanover after riding southward on Abbottstown Street. He had already ordered the 1st West Virginia and 1st Vermont back toward the fighting. When he rode into the center of town, he rallied some of the disorganized companies of the 18th Pennsylvania. One Pennsylvanian recalled that the general formed the men in line of battle and gave orders "not to shoot until they could see the color of the Rebel's eyes." After reforming, these elements of the 18th "marched down Frederick Street" to the support of the 5th New York, whose attack had stalled.[129]

After he passed through the town, Farnsworth continued to seize the initiative. According to Major Hammond, the Union troops regained the offensive at this time and once again began to push the Confederates southward. Hammond reported that when Farnsworth arrived, the 5th New York was "reformed and made another charge driving the rebels in confusion along the road and through the fields."[130] Many Southern soldiers retreated out on the Westminster Road, while others made their way back through various crop fields, fence gaps, and hedge openings.

During the counterattack, the actions of one Union soldier were enough to merit a Medal of Honor. Like many occasions when this award was given in the Civil War, this incident involved the capture of an enemy regimental color. Pvt. Thomas Burke of the 5th New York described how he seized the flag of the 13th Virginia. According to Burke, after the 5th New York counterattack "...we saw men scattering in every direction. As we neared the battery which was still being served, I noticed a Confederate flag and started after it just as Corporal Rickey did the same thing. Two mounted [men] were in charge of the colors and it was a race. Rickey had gone 200 yards perhaps,

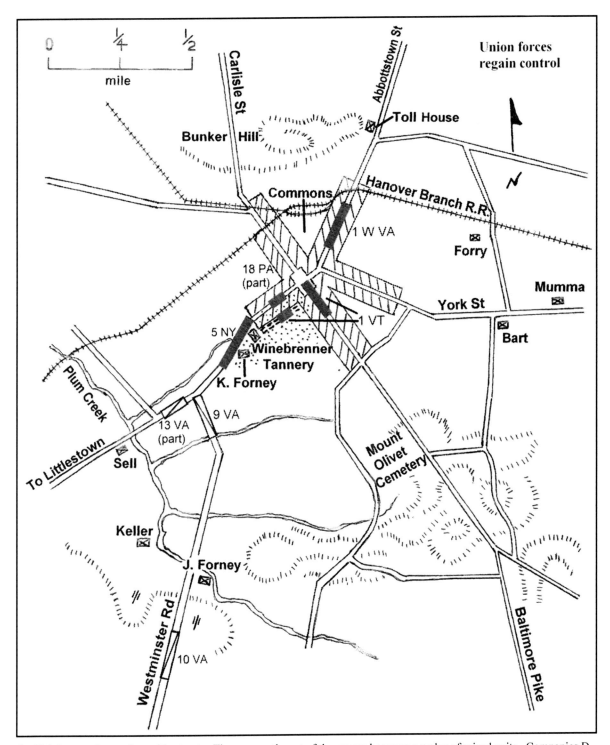

1- Fighting continues along side streets. The area southwest of the square becomes a melee of mixed units. Companies D, M (1st VT) and elements of 18th PA support the 5th NY. **2-** 13th VA and 2nd NC fall back along roads and through fields as 5th NY presses the counterattack. The 5th stalls upon receiving fire from the Confederate "reserve", likely the 9th VA. The New Yorkers regroup, then continue to push the Confederates back to the high ground south of town. **3-** 1st W VA races back to the town and takes position on Baltimore Street.

when his horse was shot and thus I was left to go it alone. Meanwhile the firing was sharp from both sides; but I gained on my prize and closing in on the men, as I used my carbine with good effect, I called on them to surrender. My command was almost instantly obeyed and I disarmed each man of carbine, sword, and pistol, after which, I rushed them ahead of me as fast as our horses - and they were very tired - would take us back to our lines."[131] Burke's Medal of Honor was the first awarded in Pennsylvania during the campaign.

Postwar photo of Col. Richard Beale, 9thVa. (G & B)

At least one Confederate officer had a different perspective than Major Hammond on this portion of the battle. Col. Richard Beale of the 9th Virginia asserted that the fire of his regiment broke the momentum of the Union counterattacks, and caused the Northern men to fall back toward Hanover. According to Beale, the Union mounted troops retreated from this area first, while the Southern forces still held the ground southwest of town.[132]

Whichever was the case, adversaries began to disengage, with the Northern men falling back toward Hanover and many Southern troops retreated behind the higher ground south of town. It became apparent to both sides that further mounted movements along the Hanover-Littlestown Road would not be successful. After the initial Confederate assaults, much of their unit cohesion had been lost for the time being. Since the 10th Virginia was then the only Southern regiment on the field that had not been heavily engaged, Stuart did not have enough reserves present at that time to launch another assault. On the Union side, Farnsworth then had most of his brigade available but faced some troubling terrain considerations. The hills south of town masked the movements of any approaching Confederates and also offered cover and concealment for the Southern men who had retreated to that area. Some of this higher ground was almost invulnerable to attack, and any potential Union offensive movement had to be considered and planned carefully. At this point, what could be called a "mutual standoff" began to evolve. The mounted troops of both sides began to fall back toward positions of greater security, where they could regroup.

Meanwhile, dismounted skirmishers began to replace the mounted soldiers in the area south of town. One Virginian described the Union skirmish line that advanced from the direction of Hanover as "extending across our front and as far to the right as we could see."[133] The nature of the battle was beginning to change, and a whole new phase of fighting was about to begin.

When the battle started, General Kilpatrick was at the head of his division near Abbottstown. Upon hearing the first cannon shots, he immediately spurred his horse into action and galloped toward the sound of the guns. Samuel Gillespie, of the division headquarters guard, believed this ride lasted about twenty minutes. Gillespie also noted that "several of our horses were injured in trying to keep up with him as the column opened ranks to let us pass."[134]

Kilpatrick returned to Hanover just after the heaviest fighting had subsided.[135] The overall situation, however, was still unsettled as dismounted troops continued to skirmish south of town. The general rode through the town to assess the state of affairs along the front lines. Sometime after his arrival, he also established a headquarters in the Central Hotel on the Hanover Square. This site was centrally located with road access in multiple directions, and the upper floor windows gave a view of much of the surrounding area.[136]

Sometime later, a captured Confederate officer was escorted to the Union headquarters. Lt. Col. William Payne, 2nd North Carolina, was the highest ranking soldier taken prisoner that day. After his horse had been killed, Payne made his way into the Winebrenner tannery then managed to fall into a tan vat, completely immersing himself. A wounded Union soldier who had also taken shelter in the area then helped him out of the vat. They both watched the fighting taking place around them

Market House and Central Hotel, post 1860 view. Thomas McCausland was the proprietor of the hotel during the Civil War. The federal census of 1860 cites eighteen individuals who resided in the hotel at that time. (GPL)

for a time. After what Payne claimed to be "an hour or more," the two came to the conclusion that the battle had ended. At about that time, another Union man rode up and demanded that the Lieutenant Colonel surrender; whereupon, he was escorted to Kilpatrick's headquarters.[137]

Payne was also the highest ranking officer wounded in the battle. He was struck across the back of the neck in the melee before his capture. Apparently the wound was not serious, although he stated that it caused him great pain when he fell into the tan vat.[138] The Confederates had at least one officer killed that day: Capt. James Dickinson, Company K, 10th Virginia.[139]

The 2nd North Carolina had seen more than its share of hardship by that time. After several actions in Virginia, the attrition during the movements through Maryland and Pennsylvania, and the fighting at Hanover, the regiment had been decimated. A number of Payne's men were captured that day. One of particular note was James B. W. Foster from Northampton County, North Carolina. Foster had enlisted in July of 1861 at the age of 59. By the time he was taken prisoner at Hanover, he was (most likely) 61 years old.[140]

Less than a half mile from the scene of Payne's tan vat adventure, Stuart himself barely eluded capture. When the general reached the field, he rode forward with some of his staff members to assess the situation. Accounts seem to indicate that at this point, he was positioned on the Hanover-Littlestown Road, west of the Karle Forney farm site. After the battle began to turn against his forces, it was not long before retreating troops were rushing past Stuart and his aides. According to Capt. W. W. Blackford, the Confederate officers soon "found themselves at the head of the enemy's charging column." Stuart guided his mare, Virginia, over a low hedge along the road and then began to gallop through fields south of the Hanover-Littlestown Road, pursued by a "party of twenty-five or thirty men." Several of the Confederate officers raced their horses through a field of "tall timothy grass" with their pursuers "firing as fast as they could cock their pistols." What they could not see until they were almost upon it was a "gully fifteen feet wide and as many deep stretched across their path." Blackford claimed he was riding side by side with Stuart, and as his horse made the leap he turned his head to see how Stuart had fared. "I will never forget the glimpse I then saw of this beautiful animal away up in mid-air over the chasm and Stuart's fine figure sitting erect and firm in the saddle."[141]

Maj. H. B. McClellan stated that although some of Stuart's staff successfully made the leap, others "landed their riders in the shallow water, whence by energetic scrambling they reached the safe side of the stream. The ludicrousness of the situation, notwithstanding the peril, was the source of much merriment at the expense of these unfortunate ones."[142] After this chase, Stuart raced back to the higher ground south of town. There he began to consolidate and redeploy Chambliss's Brigade on both sides of the Westminster Road and prepare for further developments.

By the end of the day, the battle had left its mark upon the area southwest of Hanover. One local man counted twenty-seven dead horses along the Westminster Road, from its intersection with Hanover-Littlestown Road to the Samuel Keller farm.[143]

In the town, many of the civilians recovered from the initial shock and began to give aid to the casualties. Some residents tended to wounded men lying in the streets. In other cases, soldiers made their way or were taken into the closest shelter that could be found. Various public buildings, and even homes, were turned into impromptu hospital sites. Several soldiers were particularly impressed with the conduct of the women of Hanover. Lt. S. J. Jackman (Company E, 18th Pennsylvania) said, "I saw ladies almost carrying wounded soldiers into their houses. We couldn't keep them off the street. God bless the women. They were brave and good in that town."[144] Lt. J. P. Allum (Company B, 1st West Virginia) was also inspired by the courage of the local ladies. After the war, he stated, "I have often said that the women of that place were the bravest that I ever saw. They stood in the doors of their houses and waved handkerchiefs and cheered us as we charged through the town, although the bullets were flying thick and fast."[145]

In the Hanover square, Reverend Zieber ministered to a wounded Confederate, Sgt. Isaac Peale. This North Carolinian had not only been shot in the chest, but he also fractured his skull when he fell from his horse. Although blood was "issuing profusely from his head," Peale had partially recovered consciousness and asked the pastor to pray for him, even as Zieber washed the blood from his face. At about that time, several Confederates raced across the square; one officer jumped off his horse, raised his sword, and demanded to know what Zieber was doing to the wounded man. When the pastor assured him that he was aiding the soldier and that the Confederate would be well cared for, the Southern officer thanked him, then mounted and rode away.[146]

Before long, Peale was taken to a hospital site. At one point, he recovered consciousness and spoke; it was then discovered that he was Catholic. A message was then sent to Father John B. Catani at the Conewago Chapel, who arrived and delivered the last rites. The next day, Peale died, and his body was buried at the Catholic cemetery near the chapel, then later exhumed and reburied in the South.[147]

On the outskirts of town, a few wounded soldiers made their way into the Karle Forney home. When brothers Samuel and John returned to the house, four wounded men were lying in the front room including a Confederate named Samuel Reddick. This Southern man had been shot in the chest and realized he was close to death. At some point, he pulled a copy of the New Testament out of his pocket and revealed to one of the Forney girls that his sister's address was written on the inside. The next day he passed away under the care of the Forney family and was buried near the southwestern edge of Hanover. Shortly after the battle, Miss Forney began to exchange letters with the soldier's family. Eventually the body was exhumed by acquaintances and taken to North Carolina. (Although a North Carolinian, Reddick actually served in the 13th Virginia. In some secondary works, he has been mistakenly listed as a 2nd North Carolina soldier.)[148]

At least one local had decided to fight back. S. J. Mason of the 5th New York recalled "an old man, with a head as white as snow, coming out on the balcony of his house and firing a double-barreled shot gun into the rebel ranks, as we were charging past his house."[149] General Kilpatrick's official report makes the observation that [later that day] "the main streets were barricaded and held by our troops and the citizens, who gallantly volunteered to defend their homes."[150]

The Confederates had a decidedly different take on the conduct of Hanoverians that day. Captain Blackford observed some of the action from the southwest of town. He recalled that "the long charge in and the repulse out and the hot skirmish fire opened upon them from the windows on the streets by citizens had thrown them into utter confusion..."[151] Surgeon Archibald Atkinson of the 10th Virginia was more blunt: "At Hanover the citizens fired at us from the windows. Why the town was not burnt I do not know."[152]

CHAPTER FIVE
STANDOFF

The Southern forces had initially done well. Chambliss's aggressive action had smashed much of the 18th Pennsylvania in short order. Lieutenant Colonel Payne, in particular, had led a tremendous assault with his 2nd North Carolina and probably accomplished as much as humanly possible under the circumstances. But Chambliss's Brigade had not been able to resist the greater Union numbers which had reassembled after the first shots were fired. Probably less than one half-hour after the outbreak of fighting, the town was once again under Union control.[153]

At this point, Stuart faced very limited options. Union forces were gaining strength, while Lee and Hampton had not yet arrived. There was no way Chambliss's men alone could regain control of the road through Hanover that Stuart had intended to use. Certainly a withdrawal at this particular time could not take place without leaving Fitz Lee and Hampton isolated. A standoff ensued in the late morning hours as Stuart waited for his own remaining forces to arrive. By then, Chambliss's four regiments controlled much of the high ground for several hundred yards on both sides of the Westminster Road.

In the meantime, the horse gathering details from Chambliss's Brigade made their way back to the Confederate positions. Capt. William Graham's detachment of the 2nd North Carolina had seized about twenty animals since it had left the main body. More to his surprise, he also captured a Northern soldier well away from the battlefield. The Southern captain had traveled about "three miles from the command" and was waiting in a road with another cavalryman and the captured horses. About that time, a Union officer dashed out of a patch of woods several yards in front of the two Confederates. Graham immediately pointed his pistol at the Northern man and demanded his surrender. The Union officer turned over his weapons; on the man's sword was the inscription "Lt. Shields 18th Penn. Cav. presented by the ladies of Alleghany City." The lieutenant and about twenty-five men had been on the right flank of the column and had been overrun. Apparently these Pennsylvanians had been dispersed over a wide area in the confusion that followed. Shields told the two Southerners that some of his Union comrades were "scattered" in a nearby woods. Graham stated that at that time "I signaled my men in and we struck out for the pike. There were only six of us. It was one of the most unpleasant rides I ever took. The probability was that at any time a squad of the 18th Pennsylvania, too numerous for us, would overhaul us and captors would become prisoners. I however, got in with my horses and prisoner and sent the latter on to the provost guard."[154]

On the Union side, Farnsworth, in his first battle as a general, had handled his brigade decisively in reacting to the crisis. Other officers, particularly Major Hammond, had also performed admirably. With the arrival of Kilpatrick, the Third Division then had its highest ranking officer on the field, and the command transition was complete.

General Kilpatrick had a lot to evaluate. He had technically been placed in command of the division on June 28. But for all practical purposes, June 30 was the first day all his troops were actually under his direct supervision. He had already started to assess the situation and had taken steps to solidify the Union positions. Probably even before his headquarters was established, the general began to receive various reports concerning the condition of the regiments of Farnsworth's brigade. Although these troops held the town, Stuart's men maintained a threatening presence on higher ground overlooking Hanover. Kilpatrick was faced with the possibility of another Confederate assault at any time, and his first priority was to make sure Farnsworth's brigade was positioned to defend against this potential attack.

Most of Farnsworth's men were positioned in the streets of Hanover at this point, but in the meantime, a whole new phase of the battle had evolved south of town. Much of the farmland between the Hanover-Littlestown Road and Baltimore Street remained in dispute for several hours as dismounted skirmishers fought through this area. When cavalry fought on foot, it was a common practice for one of every four men to hold the reins of horses, while the other three men advanced. It is likely that most, if not all, of the companies involved here utilized that tactic. Although this fighting did not have the intensity of the violent, mounted charges, it was of longer duration. Steady shooting continued, more or less, throughout almost the entire day.

Some of the 18th Pennsylvania had survived the initial fighting relatively intact. Part of the regiment occupied the town, and barricades were built across a few streets. Other companies skirmished "in the suburbs dismounted" throughout the morning and afternoon.[155] As the day progressed, scattered elements of the 18th continued to rejoin their unit. Regarding the 5th New York's position, Major Hammond reported that skirmishers were sent forward "and a reserve force placed at the outer edge of town."[156] Most of the skirmishing by the Pennsylvanians and New Yorkers took place southwest of Hanover, in the area between Frederick Street and Baltimore Street.

The 1st West Virginia continued to hold much of the southern portion of town. When the regiment had raced back through Hanover, it had taken a position parallel with Baltimore Street and faced west to confront Chambliss's Brigade. This deployment, however, was changed when Hampton's Brigade later appeared in force on the high ground near the cemetery. At that point, at least some of the 1st West Virginia were redeployed to confront this new threat and to avoid enfilade artillery fire.[157]

Meanwhile, one battalion of the 1st Vermont (four companies), under command of Maj. John Bennett, was positioned east of Hanover.[158] This battalion included Companies D and M, which had been involved in the counterattacks. Bennett's men were deployed to guard the Union left flank and kept an active presence along the axis of the York Road. As the day wore on, they were involved in lively skirmishing as the buildup of Confederate troops increased the pressure on the Union left.

Just north of town, Elder's battery continued to fire from Bunker Hill. After the Union counterattacks, a portion of the 5th New York was ordered to support Elder's artillery.[159] Apparently, the New York men did not remain in that capacity for long, possibly because two battalions of the 1st Vermont were also assigned to the same role. (In the midst of battle, it was not unusual for coincidental and sometimes conflicting orders to be given by different officers. A Vermont soldier stated that Kilpatrick personally ordered men of his regiment to support Elder's guns. Farnsworth may have instructed the 5th New York to support the battery at about that same time.)[160]

Soon after Kilpatrick raced back to Hanover, units of Custer's brigade began to reach the battlefield. The 7th Michigan was the first of Custer's command to arrive; the 1st Michigan and Pennington's battery arrived shortly after. Their initial position was likely somewhere between Abbottstown and York Streets. Two squadrons (four companies) of the 7th were ordered to dismount and "advanced as skirmishers under the command of Lt. Col. [Allyne] Litchfield" while the rest of the regiment remained east of Hanover for a short time.[161] At this point, Pennington's six three-inch rifled guns had not yet fired. They would, however, soon be in action from another area.

As the positions of Union regiments became more solidified, Kilpatrick began to deal with other tactical matters. By this time, the Union left flank was relatively secure. But Chambliss's Brigade still posed a threat from the high ground southwest of Hanover. Another critical factor was control of the Hanover-Littlestown Road. This route was Kilpatrick's line of communications with the Army of the Potomac's headquarters, which was at Taneytown, Maryland at that time and also the Union

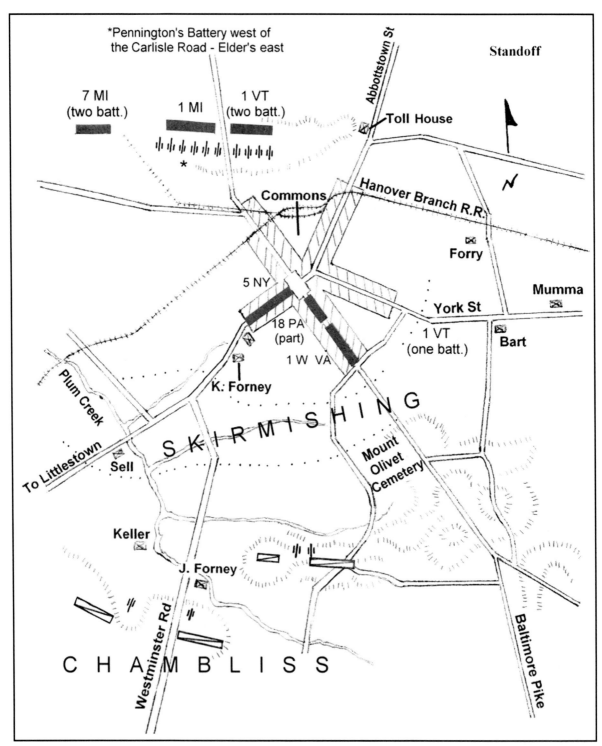

1- Mounted forces disengage. Union control of the town is consolidated as 1st and 7th MI, and Pennington's battery arrive. These units are initially positioned between Abbottstown and York Streets. Shortly after, they are ordered to higher ground north of town.
2- Chambliss's brigade regroups on hills south of town. The regimental positions are conjectural.
3- Artillery fire continues while dismounted skirmishers are engaged.

Twelfth Corps, which was approaching Littlestown. Chambliss's men were astride that line of communication.

Kilpatrick also knew from an intercepted dispatch that Fitz Lee's Brigade was approaching from the southwest and, at that very moment, might actually be on the Hanover-Littlestown Road. Also of importance were the whereabouts of Custer's other two regiments. The 5th and 6th Michigan had not yet been heard from in the direction of Littlestown. There was a possibility that these troops might be cut off from the rest of the division as they moved toward Hanover. Kilpatrick's decisions at this point reflected all these concerns. He did not allow the 1st and 7th Michigan and Pennington's battery to remain east of Hanover; before long, they were shifted to Bunker Hill. According to Pennington, his men were in the first position for only "about a quarter of an hour." Then "the command was moved to the right of town and a new line of battle taken up. A battery of the enemy soon opened, to which I replied."[162]

Because of these movements, increasing numbers of Federal cavalry were deployed near Bunker Hill. The two battalions of the 1st Vermont remained in support of Elder. Now the 1st Michigan and two battalions of the 7th performed that same duty for Pennington. But the importance of the position went well beyond that of just being an artillery platform. The hill provided a staging area from which any reserve could be dispatched to counter a Confederate flanking movement or sent south of town to support other Union troops. As a final resort, it could also be used as a rallying point should Stuart launch another attack and try to regain control of Hanover. Throughout the day, an active Union presence was maintained in that area. Some of the Union guns were planted on land owned by John Bair, who had four acres of wheat and two acres of oats trampled and destroyed. Bair also lost eighty panels of his post and rail fence, which had been torn down to make way for the batteries; the wood was later burned for cooking fires.[163]

Throughout the day, interaction continued between the civilians and Union troops. Certain businesses were bound to attract attention; one was the blacksmith and carriage-making shop of Alfred Michael, situated on Carlisle Street. Second Lieutenant Winchester Dodge, Company F, 7th Michigan swapped his horse for one of Michael's. The officer left a receipt, which stated that his animal was suffering from a sore back. After the battle, Federal officers pressed this shop into service to perform blacksmith work for the Union army. Conrad and Lewis Tate, Michael's employees, "shod Union horses from morning until night for at least seven weeks."[164] Their work would keep many Federal horses in a fit condition for the upcoming campaigns.

CHAPTER SIX
ARTILLERY

Great care was taken by Civil War artillerymen to properly pack and secure artillery rounds within the ammunition chests before any troop movement. These precautions were critical to prevent the sudden shifting of projectiles while the wagons were pulled over roads that were often not well maintained. Movements through other uneven terrain features were often necessary including rocks, creek beds, hills, stumps, etc. It was not unheard of for an accidental detonation to occur as the caissons and limber wagons sustained a sudden jolt.

The first, and apparently only, artilleryman to lose his life that day was mortally wounded in a dramatic incident that occurred well away from Hanover. As the leading Union troops approached Abbottstown that morning, an ammunition chest exploded and fatally injured James Moran of Battery M (Pennington's), 2nd U. S. Artillery. One local boy, Josiah Thoman, gave an account of this incident, which occurred just south of Abbottstown, near the mill owned by David Hollinger. Thoman stated that on the morning of the battle, he was working in a field east of the Abbottstown Road when he

> ... heard an unusual noise on the Hanover pike, as of the rumbling of heavy wagons. We hastened to the highest point on the farm, from where we could see across to the pike at a point about one-half mile south of Abbottstown, or near what is now Hartman's mill, at the time either Hollinger's or Arnold's, formerly Bender's Mill.
>
> Well on the pike we saw an army of cavalry and army teams traveling at a rapid gait. The pike seemed to be full of soldiers, as far as we could see going towards Abbottstown. All at once there was a very heavy report, as if from a cannon; so we thought they are fighting; but suddenly there was a halt of the whole force. Then at once they turned back toward Hanover and the pike was soon cleared and everything quiet, excepting we could hear cannonading in the direction of Hanover. About noon or soon after we went over to the pike, to see what had happened. We were told that when the army passed down between the house and barn on the farm at Hartman's mill, the pike being rough and as an artillery caisson or powder wagon was rapidly going down a little hill, the jalling[sic] exploded the shells and the caisson was blown to pieces. The rims of the wheels were lying there in the road with several dead horses, while the driver was lying in Squire John Elder's, (now Joseph Elder's residence) with one leg amputated and in a dying condition. He died that afternoon or night and to the best of my recollection was buried at the Catholic Church, Paradise. One soldier remained there with him until he died.[165]

All the cannoneers at Hanover were members of horse artillery batteries. One distinguishing factor of these units was that each member of the gun crew actually rode his own horse. (By comparison, members of batteries attached to infantry forces walked or sat on wagons while horses pulled the cannon.) The horse artillery allowed greater mobility; the mounted men could keep pace with the movements of the cavalry units they were intended to support.

A typical Confederate horse artillery battery contained four guns. In Stuart's command, these batteries were not formally attached to specific cavalry brigades but were utilized as necessary to support the mounted troops. Since this expedition required mobility, a minimal number of cannon were taken, while the remainder of the battalion was assigned to other cavalry brigades. In his official report of the campaign, Stuart stated that at the start of the movement, "We had no wagons or vehicles excepting six pieces of artillery and caissons and ambulances."[166]

These six Confederate guns belonged to the batteries commanded by Capt. James Breathed and Capt. William McGregor.[167] Although many of their men were from Virginia, these units also contained individuals from other areas of the South. Breathed and McGregor were natives of Maryland and Alabama, respectively, and both batteries also had enlisted some soldiers from those two states.[168] Like their Union counterparts, these gunners were part of experienced, battle-tested units. They were highly regarded by not only the cavalrymen they served with, but by their Union adversaries as well.

Historians have generally held that in the summer of 1863, Breathed's battery had four three-inch rifled guns, while McGregor command was composed of two Napoleons (twelve-pounders) and two three-inch rifled pieces. But evidence given by two officers indicates that McGregor's guns were more varied than is often supposed. In a postwar letter, Capt. Wilmer Brown referred to the battery having a Blakely rifle during the Pennsylvania campaign.[169] The battery commander himself in another letter confirmed this statement. McGregor wrote that "Captain Brown is correct about my Blakely being injured; the elevating screw having been broken at Hanover and Carlisle... The gun could be used after the screw was broken, as we often did; but the firing was not accurate." He also mentioned the difficulties in acquiring shells for the piece because "the Blakely ammunition was different from that of my other guns."[170]

Normally, Breathed's and McGregor's batteries combined would have totaled eight artillery pieces, but one of these units was without its full complement during Stuart's expedition. In a letter written shortly after the battle, Lt. Francis Wigfall of Breathed's battery stated, "Our battery and two guns of McGregor's were with the cavalry... on the expedition round the enemy."[171]

On June 30, Henry Sell witnessed some of the battle from his farm along the Hanover-Littlestown Road. According to Sell, "A Confederate cannon planted on the Keller farm to the south fired the first shot of the battle."[172] From Keller's land, this artillery provided support for Chambliss's assaults in the early stages of the fighting. Evidence suggests that the two-gun section of McGregor's battery fired those first artillery shots.[173] (See Appendix E for a more detailed discussion on the actual Southern artillery positions.)

McGregor's guns, however, were not the only cannon that traveled near the front of the column that morning; a two-gun section of Breathed's battery also moved with Chambliss's Brigade. (Meanwhile the other section of Breathed's battery was with the rear of the Confederate column.)[174] According to one of Breathed's gunners, Henry Matthews, his unit was near the action when the Confederates charged into the town, but apparently had not fired at this point. Matthews stated, "When the N. C. regiment was driven back, Breathed wheeled his guns to the rear and galloped his battery back to a hill on the left of the road (coming from Hanover) overlooking the town. We had not long to wait before the enemy threw out skirmishers, advancing through a wheat field outside of the town and to the right of the road. Immediately in front of these Yankee skirmishers was a stream running through the meadow, having banks about ten feet high. We soon got the range of this stream and threw over shells in such a way that it was made very uncomfortable for those blue coats."[175]

Since Confederate artillery also targeted Union troops in the town, several houses were struck. The experience of the Henry Winebrenner family was a somewhat unnerving one. While the family was taking shelter in the cellar, a shell fired from the southwest of Hanover tore through a balcony door on the second floor of their Frederick Street home. The projectile then smashed part of a bureau, continued through the floor and ceiling, and struck a wall on the lower level of the house.[176]

By the time of the Gettysburg Campaign, the horse artillery in the Army of the Potomac was composed of nine separate batteries. The two assigned to Kilpatrick's division were veteran units commanded by highly respected officers. Battery E, 4th U. S. Artillery, was commanded by Lt. Samuel Elder, a native of Harrisburg, Pennsylvania. Battery M, 2nd U. S. Artillery, was commanded by Lt. Alexander Pennington, a native of Newark, New Jersey.[177]

These batteries had served in the United States Army before the Civil War. Each had spent much of the 1850's in various frontier locations. Partly because of their lineage in the pre-war army, these

Lt. Alexander Pennington
(LOC)

commands had a much higher percentage of foreign born soldiers than the typical state units. In Pennington's battery, around sixty percent of the enlisted men who served in the Civil War were born outside of United States soil. The battery also had a number of native Ohioans, Pennsylvanians, and New Yorkers.[178]

Both of these units were composed of three-inch ordnance rifles, although Elder's battery had four guns while Pennington had six. Elder reached Bunker Hill first; his cannon were in action shortly after the first shots had been fired. But Pennington's arrival on Bunker Hill precipitated a renewed burst of activity on both sides. According to Pennington, when his battery reached its new position, it immediately came under fire from Confederate guns; his own men promptly responded. He stated that this intense fire lasted about "twenty minutes" before subsiding.[179]

The cannon fire then became more sporadic, at least for a time. (Later, the intensity of the artillery duel increased yet again.) Casualty figures suggest that for most of June 30, the primary targets of both sides were opposing cavalry, much more than artillery. By the end of the day, several cavalrymen had been wounded or killed by artillery rounds, but the artillery units were almost completely unscathed. Other than James Moran, apparently no men were killed in either Elder's or Pennington's units. On the Confederate side, the only June 30 casualties listed for the batteries of Breathed and McGregor were several men captured near Westminster, Maryland. These soldiers were taken prisoner by Union cavalry under Gen. David Gregg, who passed through that area after Stuart's column had departed.[180]

In the meantime, it remained uncertain who would control the area between the main lines; for several hours, dismounted skirmishers contested much of the farmland south of town. After the violent, mounted attacks, Kilpatrick and Stuart evaluated the developments and planned their next moves. Meanwhile, other dramas were about to unfold as more units made their way toward Hanover.

CHAPTER SEVEN
LEE'S SCREENING MOVEMENTS

While Stuart's main body moved northward on the Westminster Road, Fitz Lee's Brigade screened the column by riding along roads to the left of Chambliss and Hampton. Lee's Brigade, approximately 2,100 men, was composed of five regiments, the 1st, 2nd, 3rd, 4th, and 5th Virginia. His flanking force was to provide an early warning and defense against any potential Union attack from the direction of Littlestown, while it also scouted and gathered information on Union movements. The back roads necessary for this screening movement amounted to around a thirteen-mile route, but the distance was only part of the difficulties. Since Stuart had originally intended to avoid Union cavalry, Lee needed a route that did not bring his brigade too close to Littlestown. Considering that Lee was in enemy territory and had never traveled these roads, this was no easy task, and he likely used at least one local citizen as a guide. The memoirs of one soldier seem to indicate that the 1st Virginia was the leading regiment in the march that morning.[181]

Fitz Lee's troopers began their screening movement from the Union Mills area by initially proceeding to the northwest along the Baltimore-Littlestown Pike. At that time, moving south along the same road were brothers Solomon and William Snyder and Nathaniel and John Waltman. These four Adams County men were possibly the first Pennsylvanians to encounter Confederate cavalry that day. Like other locals, they had learned that the Army of Northern Virginia had moved into Pennsylvania. Since the known presence of Confederate infantry at this point was much greater to the west and north, they decided to move their horses south to get them away from the perceived threat.[182] The irony could not have been greater since they were now headed directly toward Fitz Lee's Brigade. Both Snyder brothers later testified that they were four or five miles south of Littlestown on the Baltimore Pike, when they "saw a body of Rebel Cavalry coming up the road" who then took a bay horse from John Waltman.[183] Shortly after, the Confederates seized Snyder's five horses plus another of Waltman's. If the estimation of distance is correct, this incident occurred well south of the Maryland line.

At some point, Lee's men needed to turn east off the pike to avoid Union troops near Littlestown and still screen Stuart's main column. Most likely, this first turn occurred onto either Humbert Schoolhouse or Mathias Road.[184] Now the presence of a local guide became critical; the troops were moving through hostile, unknown country, in an area where there was no simple, direct way to reach their intended destination.

While the Confederates rode northward, George Bair was working the fields of his Union Township farm. Bair resided between Line and Clouser Roads, several hundred yards north of the Pennsylvania state line. Like the Snyders and Waltmans, this farmer's day was about to take an unexpected turn. In fact, Bair got the day off work when soldiers seized the bay mare directly from his plow. Meanwhile, a black mare was taken from his stable.[185]

Regardless of the route taken to reach the Bair farm, Lee's force continued northeast, using Clouser Road, then Barts Church Road. As they moved through Union Township, Adams County, more opportunities to acquire fresh horses were presented. About three-fourth mile from Bair's residence, Confederates took a horse from the stable of Philip Fase, whose farm was located near the Barts Church Road/Pine Grove Road intersection.[186] (Fase was apparently not home at the time, but he would have his own encounter with Lee's men before long.) The brigade then turned onto the Pine Grove Road, which, for a short time, became their main artery of movement. This road, and a few farm lanes to the east, saw considerable Confederate activity, which also included at least one

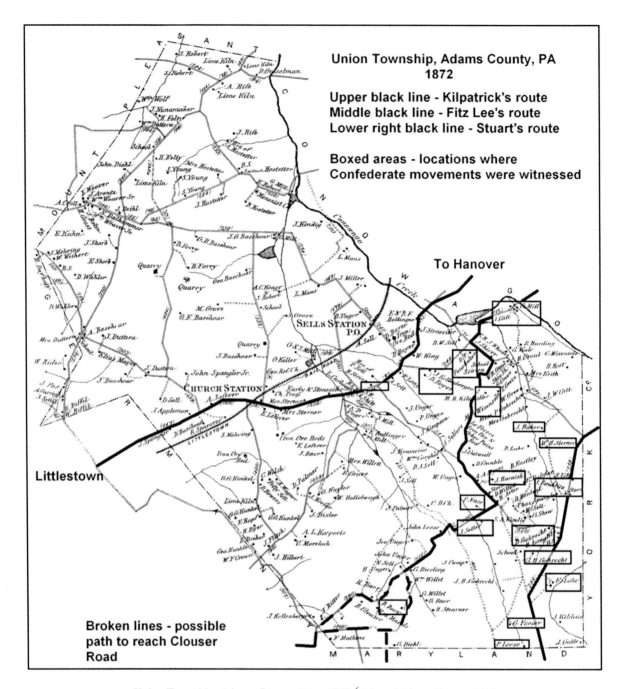

Union Township, Adams County, PA – 1872 (*Atlas of Adams County, 1872*)

Kilpatrick's route - Much of modern day route 194 still follows its original course, one exception being the area where the road crosses the Conewago Creek.
Lee's route – After reaching Union Township, Lee followed portions of the following roads: Clouser, Barts Church, Pine Grove, unnamed farm lane ("Fitz Lee Lane"), Schibert, Sheppard (past Gitt's Mill), Narrow Drive (Gitt's Mill to McSherrystown Road).
Stuart's route – After reaching Pennsylvania, Confederates travelled this approximately two mile stretch of southeastern Union Township, Adams County. They entered York County and continued northward toward Hanover.

skirmish between opposing forces. (See Chapter 8.)

Some Confederates rode the length of the Pine Grove Road to the Hanover-Littlestown Road. Near that intersection, they encountered Philip Fase driving his wagon. According to Fase, Southern soldiers removed a horse from his carriage while he was "at the Jacob Lohr mansion."[187] With these troops present on the Hanover-Littlestown Road, Kilpatrick had at least a few of his enemy positioned on his line of communications to army headquarters.[188] There would be a great amount of Confederate traffic on this thoroughfare before the day was done.

As Philip Fase's adventure was occurring, Lee's main body had already turned on to a farm lane east of Pine Grove Road. Two men who lived along this lane, Michael Kitzmiller and William Wisensale, stated that "Wisensale's sorrel horse was taken by three armed men who detached themselves from a large cavalry force."[189] John and Peter Panebaker, who resided a few hundred yards from Kitzmiller, stated that their land was "invaded by a large number of Rebels."[190]

After passing the Panebaker farm, the brigade continued on the final segment of its screening movement as it approached the Gitt's Mill area, then the Hanover-Littlestown Road. The general himself also confirmed the route taken on the last few miles of this movement. In a message to Stuart, with the heading "on march, nine a. m.," Fitz Lee stated that "The road that I am on strikes the Littletown [sic] and Hanover Road at McSherryville road [sic], one and a half mile from Hanover."[191] Lee's assessment could not have been more accurate. This intersection is almost precisely one and a half miles from where the southwestern edge of Hanover was located in 1863.

Lee had performed his task capably, moving approximately thirteen miles along back roads, while still keeping to Stuart's left. In doing so, he had screened Stuart's column from a potential attack from the direction of Littlestown. But the great majority of Kilpatrick's force had left Littlestown early that morning. Because of the distances involved, almost all of the Union troops had passed by Lee's protective screen before his brigade reached Hanover. What Lee also did not realize is that his movement would bring his men in upon the flank and rear of the 6th Michigan, which was moving along the Hanover-Littlestown Road at that time.

CHAPTER EIGHT
THE SCHWARTZ SCHOOLHOUSE FIGHT

While the majority of Kilpatrick's division was riding to Hanover that morning, the 5th and 6th Michigan were involved with their own missions of tactical importance. Company A of the 6th was sent southward from Littlestown to scout in the direction of Union Mills and Westminster. Meanwhile, the 5th guarded another road not far from Littlestown to "intercept" any Southern force that might attempt to move in their direction.[192] (Although the 5th's commanding officer did not specify which road this was, one soldier described it as running in a "similar direction" to the one on which Company A of the 6th took toward Westminster.)[193]

The rest of the 6th remained near Littlestown for a short time. Since the Union Twelfth Corps was marching that morning from Taneytown, Maryland, to Littlestown, it was critical to maintain a cavalry presence near there until the infantry got fairly close to that area. This contingent would prevent any surprise strike against the Twelfth Corps. Also of great importance was that the road from Hanover to Littlestown was the line of communications from Kilpatrick to Generals Meade and Pleasanton at army headquarters near Taneytown. Although the 5th and 6th Michigan were not present for the initial fighting at Hanover, before the day was finished, they would be involved in some very significant fighting of their own.

Col. George Gray
(GRMICHPL)

At that time, the exact location of Stuart's forces was unknown to the commander of the 6th Michigan, Col. George Gray. With Company A of the 6th and also the 5th Michigan scouting to the south, he would receive warning of any danger from that vicinity. Oddly enough, the road of greatest threat to his regiment was the one the rest of his division had taken earlier. As Fitz Lee's screening force made its way closer to the Hanover-Littlestown Road, the potential danger from that direction became a reality.

Capt. James Harvey Kidd, Company E, 6th Michigan stated that "a citizen came running in, about noon, reporting a large force about five miles out toward Hanover." In response to this news, the regiment mounted, and Colonel Gray led the men out on the Hanover Road. Kidd also related that "Several citizens, with shot guns in their hands, were seen going on foot to the flank of the column trying to keep pace with the cavalry, and apparently eager to participate in the expected battle."[194]

Without Company A, the 6th likely had about 550 soldiers on the road at that point. Colonel Gray was in a situation that might have made even the most experienced officer nervous. He led his men away from Littlestown, knowing that a Confederate force of undetermined size was likely between him and the rest of his division. This movement toward Hanover would precipitate a number of skirmishes along the axis of the Hanover-Littlestown Road. A whole new phase of fighting was about to begin.

It is likely that the movement of the 6th Michigan was made with haste once the presence of Confederates became known, particularly since no other Union troops would have impeded the progress of the regiment. After crossing the Conewago Creek, Colonel Gray's men passed the Solomon and Samuel Schwartz farms and approached the Schwartz Schoolhouse. (This schoolhouse was situated at the intersection of the Hanover-Littlestown Road and the road from McSherrystown to Gitt's Mill.) Somewhere in this vicinity, they also encountered Southern skirmishers. Since a right turn at the schoolhouse led into the rear of Stuart's lines, which were then positioned south of Hanover, this was a key intersection for the Confederates to keep an eye on. The higher ground near the schoolhouse also gave a broad view of much of the surrounding area.

As the 6th Michigan approached, these Confederate outposts began to fall back to the south of the road, toward their own supporting troops. The leading elements of Colonel Gray's regiment, still "moving in column of fours," aggressively went after these Southern skirmishers. But after moving to the right of the road and ascending a "crest," the 6th was greeted by a surprising sight.[195] A large portion of Chambliss's Brigade was now in full view, positioned southwest of Hanover on both sides of the Westminster Road. The leading troops of the Michigan column could also see a few Confederate artillery pieces in action, likely near the Westminster Road. According to Colonel Gray, as his men approached Hanover, "we came upon the enemy's skirmishers, whom we drove to their guns, which we unexpectedly found posted on our right, supported by a large force of cavalry."[196]

Before this time, an important terrain feature had kept these opposing forces out of sight to each other. This key piece of ground was a broad hill around the area of the Schwartz Schoolhouse, the highest part of which is almost two miles from the center of Hanover. (The vicinity in post war years became known locally as "Mount Pleasant.") When the 6th crossed this higher terrain and came into view, the Confederate artillerymen became aware of the threat to their own left flank. The gunners hurriedly changed the direction of their pieces and began to shell the Michigan column. The physical damage of this shelling was not severe throughout the ranks. Although Captain Kidd stated that the artillery fire wounded "several men and horses," another Michigan soldier wrote that these shots caused no damage.[197]

It quickly became apparent to Colonel Gray that his lone regiment had encountered more than he expected. Gray ordered his troops to fall back toward the Hanover-Littlestown Road, but his men were soon in for another surprise. The movement of the 6th Michigan to the right of the road brought them within striking distance of Fitz Lee's Brigade, which was then moving unseen toward their right and rear.

While the 6th Michigan had moved toward Hanover, Lee's men approached the Conewago Creek from the south. The 1st Virginia led the column past Gitt's Mill, then toward the higher ground between there and the Hanover-Littlestown Road. As the Virginians reached the crest of the ridge north of the John Shaeffer farm, the 6th Michigan came within their view. When Lee spotted the enemy, he seized the initiative and ordered the 1st Virginia to attack. According to one cavalryman, Lee "turned to the First, which was then marching in front" and said, "Charge them boys, there isn't many of them."[198] (See Appendix I for more on the location of this encounter.)

When the 6th Michigan began to fall back toward the Hanover-Littlestown Road, their officers likely thought that their withdrawal could be accomplished relatively easily. At that point, they were still several hundred yards from the Confederate artillery and the majority of Chambliss's men. But once again, a Union regiment was struck by a surprise assault, in a situation oddly similar to what had happened closer to Hanover. Within minutes, the 6th Michigan had possibly close to two thousand Confederates of Lee's Brigade charging hell bent for leather upon their right and rear. Colonel Gray stated that as his men withdrew from the first Confederate threat, "we were completely flanked by another body of the enemy's cavalry, outnumbering my command at least six to one."[199] Another Union man believed that "the rebs had got in our rear with about three thousand cavalry when they made a charge."[200]

Outnumbered and in danger of being encircled, the 6th had no choice but to get out quick. One

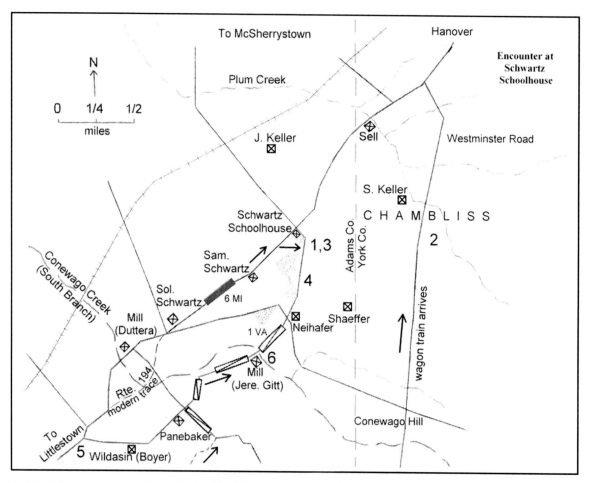

1- The 6th MI encounters Confederate skirmishers, likely west of the Schwartz Schoolhouse. After pursuing the skirmishers across high ground (Mount Pleasant), the 6th comes into view of the main body of Chambliss's brigade. The appearance of the 6th MI is a direct threat to the Confederate rear and the captured wagons.

2- Southern artillery opens fire upon the Michigan column. The 6th falls back toward the Hanover-Littlestown Road.

3- After crossing the Conewago Creek at Gitt's Mill, the 1st VA strikes the 6th MI, likely northwest of the John Shaeffer farm. Other regiments of Lee's brigade move into position and begin to outflank the 6th.

4- Companies B and F of the 6th MI fight a rear guard covering action from wooded area(s). Their positions are conjectural, but likely included the largest woods south of the Schwartz Schoolhouse. They were eventually cut off and did not rejoin their division until well after dark.

5- Much of the 6th MI regroups north of the Schwartz Schoolhouse; others race southwest with Virginians in pursuit. They soon encounter the 5th MI moving toward them along the Hanover-Littlestown Road. (The exact position of this contact is unknown; primary accounts indicate it occurred west of the Conewago Creek.) The flow of the fighting reverses as the 5th pushes Lee's men back toward Hanover.

6- Michigan troops use Gitt's Mill as a sharpshooter stronghold. From here they harass the Confederate rear. (Elements of Companies B and F of the 6th MI might have taken shelter in the mill for a time. Or, the site might have been occupied by 5th MI soldiers who screened the right flank of their regiment as it moved along the axis of the Hanover-Littlestown Road.)

Modern view looking southeast toward "Mount Pleasant" from the Hanover-Littlestown Road (Route 194), showing a portion of the 1863 Samuel Schwartz farm. It is possibly the area where the 6th Michigan crossed the "crest" en route to their encounter with Lee's Brigade. (AC)

member of the unit, John Kay, wrote, "they were surrounding us fast. (This was a mile from H.) We changed our direction and had to run as they were attacking our rear."[201] Under immense pressure, most of the regiment engaged in a running, gun-fighting retreat. Although Colonel Gray's men would eventually rejoin their division in Hanover, for the moment, they were in big trouble. The great majority of the 6th likely fell back beyond "Mount Pleasant" and then north of the Hanover-Littlestown Road. However, at least a portion of the regiment rushed back toward Littlestown with Confederates in close pursuit.

Physical casualties were not the only losses sustained here. Kay was more upset by the loss of personal items than anything else. In a letter to his parents, he wrote the following: [Some punctuation changes were made for clarity.] "...at the commencement of the fight we were drove hard by the Rebels & were ordered to throw off everything to speed our flight. I was unfortunate enough to lose everything; all the little tokens of & remembrances of Home, Affection, & Love; lost all my letters, books, what clothing I did not have on & everything. But my dear Mother's picture that is safe. Oh what a loss I have met with! I would sooner have lost hundreds of dollars than to have lost those precious mementoes of home and all its endearments. But thank God I have not lost my life yet, though the balls whistled close & the shells burst near me, He has kept me as under the shadow of his wing. I have felt it."[202]

George Bare of the 1st Virginia lost more than tokens of his family; he lost part of the family itself. Two of his brothers, Adam and Isaac, were captured during the action. Isaac was also wounded in the fighting.[203]

One of the attacking Confederates, Sgt. Benjamin J. Haden of the 1st Virginia, described the assault this way: "The charge was a race for at least *two miles*, [italics added] when they reached their command, and upon being reinforced, turned and charged us. I had just dismounted to procure the arms and horse of a Yankee who had been wounded, when Billy McClausland of my Company came back at full speed, telling me as he passed, to mount and get away as quickly as possible, as they were charging with a heavy force. I knew when he left times were getting squally, so I mounted, putting the captured horse in front of me, drew my saber and struck him a few times, and brought him out safely. We lost some good men in the charge; C. H. Koiner, of my company, had his horse killed while making the charge, and had mounted a Yankee's horse (the rider of which had been killed), was shot, the ball passing through the front of the saddle and then through him, wounding him mortally. They pursued us until we reached our support and then retired."[204] (Note: If Haden's estimation of a "two mile" race is correct, he was pursuing the troops that retreated in the direction of Littlestown. His description of the Union countercharge refers to the 5th Michigan, which was moving from Littlestown toward Hanover, as will be seen shortly.)

When the 6th Michigan came under attack, Colonel Gray "placed two companies (B and F) in position" to protect the rear of the regiment and "to check the enemy's advance."[205] Capt. Peter Weber had the unenviable task of commanding this force of possibly 120 or more troops. Weber's men utilized wooded areas for much of their covering action. Although the combat was fast, furious, and fluid, it is likely that initially they were positioned in the patches of woods located south of the schoolhouse (west of Narrow Drive).[206] This action turned into a running, gunfight but even under extreme duress, his troops maintained cohesion and discipline. They not only repulsed at least three distinct attacks but also aggressively counterattacked at least that many times.

With Weber commanding the squadron, Lt. Daniel Powers was, in effect, in charge of Company B itself. Powers was in the thick of things, in terms of both terrain and combat. His involvement in the holding action was about as eventful as possible for a soldier. According to the lieutenant, during the second mounted counterattack through a wooded area, his head struck a branch and knocked off his hat. The force of the blow threw him back on his horse "so that a volley of pistol bullets just missed" his body. In the next volley, however, a round struck his scabbard and glanced off into his horse at which point blood spurted out all over his boot. The Michigan troops maintained their momentum; shortly after, they overran and captured a few Confederates including a "big" lieutenant. In the heat of the moment, one or two of the Michigan men were so excited that they wanted to shoot the Southern officer on the spot. Powers, however, maintained his composure; he knocked down the pistols of his own men with his saber, probably saving the life of the enemy officer. Shortly after this incident, the Confederates launched another attack. Apparently at this point, the Southern lieutenant escaped in the confusion, minus his weapons, which had been seized by Powers.[207]

Another charge and countercharge followed, at which time Companies B and F fell back once again into a wooded area to regroup and reload. Any possible feeling of security they had at this point was crushed quickly, however. The Michigan men had hardly reformed when Sgt. William Keyes yelled that they were being outflanked. Lieutenant Powers was still hopeful that his company had not been spotted, but for their safety, he ordered his men "up the road a ways out of their range" while he went back to try and figure out the Confederates' intentions. He claimed he had only "…gone but a little ways when their side skirmishers came right on to me all firing at the same time." Powers' horse then received his fifth and final wound; the animal was shot in the head and killed.[208]

At this point, the Virginians commanded the Michigan lieutenant to surrender his arms. However, when Powers realized he was about to be captured, he threw his pistol into a wheatfield. When the Confederates demanded to know where the weapon was located, Powers pointed toward the wrong direction. Events could have gotten ugly at this point; one of the Southern men called him a "dammed little cuss" for lying about the pistol. Just then the "big" Confederate lieutenant, whom Powers had briefly captured, rode up and, recognizing his former captor, ordered his men to treat the

Union man with respect. Powers' actions to save the officer's life had not been forgotten.[209]

A few common themes emerge among the accounts of soldiers involved here. One was the surprise nature of the attack; another was the overwhelming numbers the Confederates were able to bring to this phase of the battle. Weber's detachment did not draw Lee's entire force after them but did occupy a significant amount of the brigade. Allen Pease of Company B asserted that the 6th Michigan was "surrounded by two brigades of rebel cavalry," and that when his men moved into the woods, they had "about three hundred rebs" specifically pursuing their detachment.[210] After the second counterattack, Lieutenant Powers peered through the woods and realized that his men were not only being outflanked but that there were "five times" as many of the enemy as his own small force.[211]

Another aspect that was typical of so many cavalry engagements was the number of horses wounded or killed. Daniel Stewart of Company B had his horse wounded in "the first fire" but the animal continued to carry him throughout the fight and retreat. He stated that after his company was separated from the regiment, they "ralied [sic] and drove the rebs three times…".[212]

Trying to fight a protective covering action had been a daunting assignment but Weber's men had achieved their objective by covering the retreat of much of the 6th Michigan. But as more Virginians rushed into the fight, Weber's detachment was in danger of being encircled. The men of Companies B and F held out as long as possible, but then raced out of the woods for their own survival. Although they had avoided being overrun, several of the Michigan men soon found themselves trapped behind enemy lines. According to Daniel Stewart, after the third attack, the detachment "…rode into a piece of woods and lay thare [sic] till near sundown…," apparently well south of the Hanover-Littlestown Road. While they hid in the woods, Confederates passed by within "forty rods" of their position but did not discover them until "near dark." Stewart said that at that point, "we had just time to mount our horses and dig out when they began to shell the woods…"[213] It was not until well after dark that these two companies were able to rejoin their regiment in Hanover.

Several landowners were affected by fighting in this vicinity. One was John Shaeffer, who resided about one-half mile south of the Schwartz Schoolhouse. Confederates had already seized four horses from his stable, but much more excitement came after the attack of the 1st Virginia upon the 6th Michigan. Shaeffer claimed that the Southern men "fought and skirmished with the U.S. forces" in that area and that during the "battles and skirmishes," the Confederates "destroyed about two acres of wheat in the ground."[214] His neighbor, Samuel Schwartz, sustained crop damage inflicted by both sides during the battle.[215] The Schwartz Schoolhouse area saw considerable activity that day.

The fighting between the 6th Michigan and Lee's Brigade can be considered part of the Battle of Hanover and as a separate action in itself. Although Stuart and Kilpatrick both knew more of their own troops were on the way, this confrontation took place without any control by either commanding general. Similar to the earlier fighting, this engagement was also completely unplanned. Although this combat did not occur in a town, references to three different towns can be found in the military records of the men who became casualties in this action. Because of the location of this phase of the battle, some soldiers' service records state that their casualties occurred near Hanover, while others cite Littlestown or McSherrystown.[216] Lt. Col. William Carter (3rd Virginia), wrote that "Our Brigade charged a party sent to meet us at Sherrystown [sic]…".[217]

It is possible Lee thought that the Confederate left was relatively safe at this point but, if so, he was in for a surprise. Another Union regiment was about to enter the fight for control of the Hanover-Littlestown Road.

5TH MICHIGAN

By late morning, the 5th Michigan completed its scouting foray south of Littlestown and started to move to rejoin the division. Two battalions (eight companies) of the regiment rode toward Hanover, while the other battalion remained to guard wagons (probably closer to Taneytown, Maryland). In the estimation of one officer, about 450 men of the regiment were actually on the road at that point.[218] The 5th, commanded by Col. Russell Alger, was also accompanied by Company A of the 6th Michigan, which had also completed its own scouting mission. After the mayhem near the Schwartz Schoolhouse subsided, the Confederates established control over a few miles of the Littlestown Road, southwest of Hanover. Meanwhile, this fighting had also spread along the axis of the road in more than one direction, as Lee's Virginians pursued at least a few of the 6th Michigan back toward Littlestown. To reach Hanover, the 5th Michigan would have to fight its way through what had become a very unstable area, as it encountered the "fallout" from the previous melee.

Col. Russell Alger
(BBGB)

As these cavalrymen moved toward Hanover, it was not long before they ran into the uproar caused by the previous fighting. The sound of shooting in front was the first sign of trouble. Then, as the leading troops of the 5th rounded a bend in the road, they saw some of the 6th Michigan galloping right toward them, with Confederates in close pursuit. According to John Allen Bigelow (a. k. a. John Allen), these members of the 6th were yelling "Rebs! Rebs!! Rebs!!!" as they raced toward Company A of the 5th. These fleeing men rushed into Company A's ranks with Virginians chasing after them.[219]

Apparently, both sides were caught by surprise at this point. The Southern men desperately tried to halt their momentum as they realized they now faced a considerable Union force that was definitely not retreating. Bigelow claimed that the closest Confederates became "crowded into a mass by those in their rear, 100 feet in our front."[220]

Confusion reigned as the leading elements of the columns made contact. A wild melee ensued, similar to the earlier fighting at Hanover. Pistols and sabers were used freely by at least a few mounted soldiers at the front of the column. Meanwhile, several men from Companies A and F of the 5th Michigan dismounted with their Spencer rifles and rushed into the fields along the road. Shooting became more general as the men on foot attempted to gain a clear line of fire toward the Confederates. The combat quickly evolved into a running gunfight as the momentum shifted, and the Virginians began to fall back toward Hanover.[221]

Although it is not certain exactly where these two units first made contact, this combat originated west of where the Conewago Creek crosses the Hanover-Littlestown Road (i.e., in Union Township, Adams County). Some 5th Michigan men made a clear distinction between this action and the fight at Hanover, which they considered to be a completely separate engagement. Maj. Luther Trowbridge referred to this action as "our skirmish at Littlestown."[222]

Several Adams County civilians were affected by this phase of the fighting. One was David Boyer, who owned more than 130 acres of land just west of the Conewago Creek, mostly within sight of the Hanover-Littlestown Road. He and William Wisensale witnessed this encounter but greatly overestimated the number of troops; they believed the Union force involved was "about 1,000 cavalry." Boyer stated the soldiers rode "back and forth" across crop fields while "there was a skirmish going on."[223] Another local man, Emanuel Wildasin, sustained extensive damage to a nine-acre field of rye "on a hill" during this same action. "Skirmishers" destroyed the grain when a "large force" of Union cavalry took position in the area.[224]

On a nearby farm, Levi Maus suffered damage to several fields of oats, corn, and wheat. He also noted that "...a number of disabled soldiers that had been hurt in the skirmishing" were present on his land in the aftermath of the fighting. A local blacksmith, Lewis Carbaugh, saw Union troops "tear down and scatter" a number of fences on Maus's land.[225]

As the 5th Michigan tried to make its way toward Hanover, some important terrain features made tactical adjustments necessary. In several areas, the Hanover-Littlestown Road is overlooked by much higher ground just to the south. Control of these hills was critical to provide safe passage for any force moving along the road itself. While much of the 5th advanced along the road, other companies of the regiment deployed in supporting positions along the high ground to their right.

When contact was made with the Virginians, the great majority of Michigan men were still on horseback. But as the fighting shifted eastward, more of the 5th Michigan dismounted; possibly a full battalion deployed on foot to the south of the road.[226] Colonel Alger gave the following description of his regiment's initial contact with Lee's troops: "... the enemy attacked me in quite a large force. I charged him, driving him some distance, dismounted my command and fought him on foot, killing and capturing quite a number. My loss was severe." The colonel also stated that since his regiment had been armed with Spencer rifles, he was "then, and afterwards, required to do very much fighting on foot."[227] As more dismounted Michigan men deployed along the axis of the Hanover-Littlestown Road, greater numbers of these weapons went into action. The presence of the Spencer was likely a critical factor in causing the Confederates to fall back.

By this time, fighting between Lee's Brigade and the 5th and 6th Michigan had taken place over a wide area. Several other residents of Adams and York County were affected by skirmishing on their land. The action that took place near Gitt's Mill provides a notable example.

In 1863, Jeremiah Gitt owned 200 acres of land in Union Township, Adams County.[228] He also operated a large mill along the south branch of the Conewago Creek, located about three-fourth of a mile south of the Schwartz Schoolhouse. Since this site was situated along the route taken by Lee's Brigade, and also less than a mile from the Westminster Road, it was bound to attract attention. Two horses were taken from his premises by what the owner described as "a large force of rebels."[229]

At some point, Union cavalrymen took shelter in the mill and used it as a sharpshooter stronghold. Gitt stated that damage to his mill was caused "by the soldiers of a *Michigan Regiment* [italics added] who occupied it and punched their guns through the glass being engaged at the time in a skirmish with the enemy."[230] Michigan troops on his land posed a serious threat to the Confederate rear. (A portion of Captain Weber's detachment could have been involved here for a time, after they were trapped behind the Confederate lines. Another possibility is that 5th Michigan soldiers fought at the mill, after making contact with the enemy west of there.[231] See Appendix J for more on the Gitt's Mill fighting.)

At least one Confederate was killed somewhere in this vicinity. While local tradition usually refers to this soldier being killed near the mill, one Hanover boy had a different notion regarding the location. William Gitt was the son of local merchant Josiah Gitt (not the previously mentioned Jeremiah). In 1869, William was in his father's store in Hanover when a visitor entered to inquire about the location of the remains of a Confederate who had been killed during the fighting. Others in the store could not help the man, even after he described the location. William then spoke up, saying,

Gitt's Mill (R. E. Spangler collection, GPL)

"A Confederate officer was killed at the blacksmith shop on Conewago Hill, near our farm, on the Westminster Road, and was buried at the barn." The visitor and a few locals then proceeded to the area and found the burial location; whereupon, these remains were exhumed and taken to the South.[232]

Conewago Hill was an important terrain feature in this area. Located a few miles south of Hanover and about one half mile east of Gitt's Mill, this high ground was critical from a logistical standpoint alone. The Westminster and Fairview Roads intersect at the apex of the hill. This intersection was important for Stuart's main column movement toward Hanover and for the Confederate retreat later that day. From Conewago Hill, Fairview also runs northwest (becoming Narrow Drive when it crosses into Adams County) and leads to the Hanover-Littlestown Road. It was crucial for the Confederates to control that means of access to prevent any Union troops from moving into the rear of the Southern lines. When the 6th and 5th Michigan moved through the area south of the Schwartz Schoolhouse, that threat was no longer only theoretical.

The protection of the captured supplies was also a factor in this area. By noon, the wagons were approaching, or had already reached, the Confederate rear. Some Southern accounts refer to Union movements that threatened the vehicles, and it was Michigan troops that caused these concerns. Capt. William Graham, after completing his horse-gathering assignment, returned at this critical stage. Graham related that "He [Stuart] came back to the wagon train about the time I reached it; the enemy were advancing in our *rear* [italics added] but Fitz Lee came from the left before they reached the wagons and drove them off..."[233] It is not certain whether any of the Michigan men knew how close they were to Stuart's prized capture. But any movement by the 5th or 6th that reached the Gitt's Mill area, or approached Conewago Hill beyond, would likely have been interpreted by Stuart as a threat to the wagon train.

After the 5th Michigan moved through the area, the situation near the Schwartz Schoolhouse and Gitt's Mill began to stabilize. Almost all of the forces were now within sight of Hanover, and officers of Lee's Brigade and the 5th and 6th Michigan reported to their commanders for further orders.

Other units had their own adventures that day along the road between Hanover and Littlestown. One small detachment of the 5th New York barely made an escape after a close scrape. Company M, commanded by Lt. Eugene Dimmick, along with Company L, commanded by Capt. Augustus Barker, were on picket duty the previous night. The next morning, while most of the division moved to Hanover, these New York troops withdrew from their outpost positions. After all their scouts reported in, the two companies began to ride along the Hanover-Littlestown Road. (This movement likely took place while the 6th Michigan was still near Littlestown.)

Even though they were detached from the division, Dimmick was not initially concerned. He stated that "It did not occur to Captain Barker or myself to throw out an advance guard or flankers, the command having so recently passed over the road; in fact, we had not the slightest idea of the enemy being in that vicinity." The first sign of trouble was the appearance of several men on high ground to the right of the road, which Dimmick termed "suspicious." When a "sergeant and half dozen men were sent to investigate," they came back with two Confederate deserters who told the New York officers that "Hampton's Brigade was only a short distance" away. At this point, the New York officers sent out an advance guard and flankers, and the movement continued toward Hanover. According to Dimmick, his men "had proceeded about three miles on our way when we were suddenly attacked by at least a squadron of Stuart's Cavalry that charged down a lane on our right flank and, for a time, matters looked pretty serious for our two small troops, of not more than forty men each, but we succeeded in holding the enemy in check by vigorous rear guard fighting and by making quite a detour to the left eventually joined our command at Hanover." In this skirmish, the New York detail sustained casualties of "several men wounded and captured with the lead horses of the detachment." Dimmick called this incident "a very narrow escape and was glad when we sighted and galloped into Hanover."[234] Once again, a body of Union troops had been caught in a surprise attack. And, once again, the Confederates had ridden down upon their right flank by a road connecting to the Hanover-Littlestown Road.

Most of the fighting southwest of Hanover was initiated by troops that made contact on, or within sight of, the Hanover-Littlestown Road. Some exceptions, however, did occur. Skirmishing occurred on a few farms near the Pine Grove Road and typified many small unit actions in the Civil War that have been lost to history. Jesse Wentz and Benedict Esale lived on two farms a few hundred yards east of this road. Both were witnesses to fighting that took place on the Wentz property. Wentz lost a horse during this action, but neither he nor Esale knew which side took the animal.[235]

The area between the main column routes had turned into a very active and "crowded" sector, as opposing patrols moved along several interconnecting roads and farm lanes. As these scouts made contact, fighting took place between detachments that were, in some cases, even smaller than company size. Whether or not historians choose to call these actions part of the "Battle of Hanover," clearly, shooting incidents were taking place over a much wider area than the immediate vicinity of Hanover itself.

CHAPTER NINE
THE TWELFTH CORPS REACHES LITTLESTOWN

During those last days of June, the task of Union cavalry was not only to gather intelligence on Confederate movements, but also to protect and screen Federal infantry. Kilpatrick's men on June 30 were in advance of the Union Twelfth Corps, which was marching from the direction of Taneytown, Maryland. This corps numbered approximately 10,000 men and its leading units reached Littlestown by about noon. The commotion caused by small detachments of cavalry to their front was noticed and commented on by several of these Union soldiers; in fact, skirmishing to the east of Littlestown was of enough concern to cause the deployment of some of these infantry units, along with some Twelfth Corps Artillery.

According to Gen. John Geary, his division of the Twelfth Corps reached Littlestown June 30 at noon, and "A half hour before reaching this place our cavalry had there a skirmish with that of the rebels. The command was hastened forward and dispositions at once made to receive the enemy, who, however, retired in the direction of Hanover."[236] Col. Charles Candy was even more specific in his report, saying that upon reaching Littlestown, "this brigade was ordered to take a position in the woods on the right of the town (Littlestown), in the direction of Hanover, and on the right of the road, and hold it at all hazards. The cavalry skirmishing with the enemy in the front, I immediately moved with the brigade to the point designated, formed in column by two battalion front, threw forward skirmishers, and picketed to my front and right."[237] It is interesting to ponder what Stuart's reaction might have been had he known how close his cavalry was to getting in a fight with Union infantry.

Like the cavalrymen the night before, these Union foot soldiers also found an overwhelming welcome from the local citizens. One infantry soldier remembered the following scene when he reached Littlestown:

> The sight which met our eyes was of the most cheering description; on each side of the main street, through which we were marching, were tables well spread with tempting looking eatables, and behind the tables were charming looking young ladies, ready to minister to the wants of the hungry soldiers. A halt was called, but the next moment three or four shots were fired on the outskirts of town, and orders were at once given to the leading regiments to "double quick" to the front to support our advance guard of cavalry. Prospects of soft bread, apple-butter, pies, cakes, and other good things vanished like a dream, as we rapidly left them in our rear. Who ate them we never knew; but as we saw no enemy, it seems doubtful whether this was not what would be considered, in the slang of the present period, a "put-up job" by the regiments which followed us. However this may have been, we lost the good things, and went into bivouac about half a mile from the town."[238]

Although this infantryman did not realize it at the time, this was no "put-up job". There actually were shots exchanged between opposing cavalry detachments not far from the town. One man involved in this skirmishing was Allen Rice, of Company C, 6th Michigan. Rice had been detached from his unit for at least the last twenty-four hours after he and four others had been ordered to gather forage. By the time this detail reached Littlestown, their regiment had already proceeded to Hanover, so the detachment continued in that direction. According to Rice, "We got about a mile and a half when we heard an awful yelling." Believing it to be Confederates, this small band wheeled their horses, attempting to flee. After racing several yards, they realized the men behind them were actually Union soldiers being pursued by Confederates. Rice wrote that "The Rebs had come round and cut of[f] the lead horses and ware [*sic*] trying to charge in to the town, but when they got up near the town a company of our men ware [*sic*] dismounted and in a field ready for them..." After this dismounted "company" unleashed a few shots, the Confederates then became the pursued and raced back in the direction of Hanover. At this point, "another Company" of Union troops chased them on horseback and "took several of them prisoners and killed some of them."[239]

The skirmishing near Littlestown caused a "ripple effect" along the Union Twelfth Corps marching column, as regiments toward the front were rushed forward for support. Further back, Joseph Lumbard of the 147th Pennsylvania saw an officer ride up to their colonel, communicate something to him, and then continue further to the rear. At this point, the soldier noted, "We were ordered to quicken our speed, and soon we noticed the troops in our rear break files to the right and left, and following their example, we discovered the aide returning with a section of Knap's battery. As soon as it had passed we closed up our ranks and made up our minds to be ready to meet the enemy at almost any time."[240]

Closer to the front of the column, one New York soldier watched as "Gens. Slocum and Geary dashed forward and soon a battery came up the road and passed on to the front, the horses on a dead run. As soon as the battery had passed the command 'fall in' was heard, and every man was up and in his place in a moment. 'Forward, double quick' was given, and the boys went tearing along after the battery."[241]

As these Twelfth Corps units rushed through Littlestown, many locals were still on the streets offering food and drink. But most of the men did not have the time to enjoy the hospitality, although some did grab handfuls of food as they passed. Lt. Robert Cruikshank of the 123rd New York remembered the scene well. He said, "The people of the town were out at their doors passing to the men as they ran by such provisions as they had in their houses, with water. Ladies waved their handkerchiefs and cheered us on. Some were in tears and some in smiles. At the hotel a number had gathered and were singing patriotic songs. If I ever felt that I wanted to fight the enemy, it was here where those ladies were calling to us to drive the Rebels back into Virginia where they belonged. Then, too, I remembered the patriotism of that state, the thousands it had fed while going to the front and the care it had given to the sick and wounded returning to their homes. This all passed through my mind and I felt that I wanted to meet them in this free, hospitable, patriotic state."[242]

The lieutenant then added, "We did not slacken our pace until we were a mile beyond the town, where we were marched into a large field, formed in a line of battle and rested. Our cavalry had run into the Rebel cavalry and had a skirmish at this place but had driven them back."[243]

Joseph Lumbard related that as the 147th Pennsylvania reached the area beyond Littlestown, "...the head of the column was turned into a field. A short distance to our right was a newly thrown up earth-work, and for it we started." Apparently some of the Pennsylvanians began to get nervous as they approached what they believed was "a fort not over six hundred yards to their front, whose deadly guns at any moment would belch forth death and destruction, and here our officers were moving us within easy range in column." As the men drew closer, however, they realized that what they had taken to be a fort was actually an "ore bank."[244] (Note: the remains of a few of these ore banks can still be seen today, south of the Hanover-Littlestown Road, near Mehring Road.)

Like the Hanoverian who had fired shots to help defend his hometown, a Littlestown resident also decided to fight. According to one Union soldier, an "old man" volunteered to show some of the Union troops where a "squad" of Southern cavalry was positioned. Not only did he guide the Union soldiers to the correct location, but in the midst of a small skirmish, he dragged one Confederate from his horse and "choked him into submission."[245]

CHAPTER TEN
HAMPTON ARRIVES

When the first shots were fired that morning, the three Confederate brigades were separated from each other and had been unable to act in any tactical coordination. Stuart stated that "owing to the great elongation of the column by reason of the 200 wagons and hilly roads, Hampton was a long way behind" when the battle started. According to Maj. H. B. McClellan, "Hampton was separated from the leading brigade by the whole train of captured wagons."[246] By the time Hampton reached the battlefield, the most intense fighting was finished.

Hampton's Brigade was composed of the 1st South Carolina, 2nd South Carolina, and 1st North Carolina regiments, along with three other units referred to as "Legions." The Cobb, Phillips, and Jeff Davis Legions were units that, in their conception, had incorporated cavalry, infantry, and artillery under a single command. But early in the war, the different arms had been separated and then affiliated with other units of the same type of service.

With the arrival of Lee and Hampton, probably before 1 P. M., Stuart continued to consolidate his hold on the higher terrain south of town. Lee's Brigade held the Confederate left and controlled much of the area from the Schwartz Schoolhouse eastward toward the Adams/York County line. Chambliss's men, positioned on both sides of the Westminster Road, controlled from there eastward through the higher ground to the Beck Mill Road.[247] Hampton's Brigade extended the Confederate line to a considerable degree. Some of his regiments were positioned in force on the Baltimore Pike, particularly near the Mount Olivet Cemetery. But his line also extended to the east of town, even to the north of the York Road. By keeping a presence near the York Road, the Confederates would be alerted of any possible excursion by Union troops to get in the Confederate rear.[248]

Hampton's control of the high ground near the Mount Olivet Cemetery was a major threat to Union troops. This elevation was bisected by the Baltimore Pike, about one mile southeast of the square, and other than the cemetery itself, it was mostly open farmland. Not only did that terrain dominate the town, but also it was much closer to Union forces than the hills where Chambliss's men were positioned. Had Stuart decided upon another attack, Confederates could have massed troops behind this ridge without Kilpatrick's knowledge.

A few Confederate artillery pieces had traveled with Hampton's Brigade that day. At least two guns of Breathed's battery (possibly all four) took position near the Baltimore Pike, in the vicinity of Mount Olivet Cemetery. One or two cannon likely fired from the road itself.[249] From there they began to shell Union troops in the town and artillery on the opposite side of Hanover. The arrival of Breathed's final two guns began the most intense phase of the artillery duel that day. Most accounts suggest that this portion of the cannonading lasted at least an hour. According to Reverend Zieber, "The screaming and bursting of shells continued for an hour or more. Several balls and shells fell on the streets of Hanover."[250]

Other than the artillery fire, the action along the axis of the Baltimore Pike was mostly long-range skirmishing. By that time, barricades had been constructed on various streets, and shooting continued throughout the fields and from some of the buildings on the southern edge of town. The 1st West Virginia lost less than ten men killed or wounded that day but had almost twice that many taken prisoner. Exactly where these men were captured is not certain. Hanover had been mostly cleared of Confederates by the time they rushed back to the fighting. But any detachment that rode well south of town might have been cut off from its regiment after Hampton's arrival. (They may have been captured in the skirmish near the Brockley farm during the Confederate withdrawal. See

Chapter Eleven for more on this rear guard skirmishing.)

Meanwhile, significant troop activity also took place to the east of Hanover. One prominent landowner in this vicinity was Samuel Mumma, whose farm sat along the York Road, about one mile east of the Hanover square. His son, also named Samuel, operated a mill nearby along Oil Creek. Mumma decided to get a better view of the excitement by climbing into a barn loft and peering out of a hole in the roof. From here he saw "many" Confederates "in possession of the ground" that was the property of a neighbor, Jacob Bart. Bart himself verified that there were large numbers of Southern soldiers around his house, but he apparently was not as eager to view the proceedings. He admitted that he hid in his cellar due to "heavy cannonading."[251] By this time, the right of Hampton's line stretched through his land and extended beyond the York Road. (Bart lived close to where the York Road intersects with the road now called Brookside Avenue.) Bart did not say whether his greatest fear was outgoing Confederate fire or incoming Union shots, but apparently his fears were not unfounded. Rev. Dr. Charles Stock also asserted that one Southern artillery piece was in the area near "Mumma's Mill."[252]

Even land several hundred yards to the north of the York Road fell under Confederate control during the engagement. George T. Forry and a few of his neighbors said that Southern soldiers had "full possession" of the ground where Forry's stable was situated and for "some distance around."[253] (Their evidence would indicate that much of the area near what is now known as Center Street was in Confederate control.)

A few different units were opposing Hampton's troops in this area. While most of the 1st Vermont was positioned in support of Union artillery, one battalion of the regiment had been detached and ordered to deploy out on the York Road. This battalion, under Maj. John Bennett, was composed of companies A, D, K, and M. These four companies held a position on the "left of the town" where they were "warmly engaged."[254] Horace Ide (of Company D) referred to this dismounted skirmishing as "pretty lively."[255] Although no major attacks occurred in this area, the skirmishers engaged in sporadic fire for the next few hours.

The Vermont men were not the only Union troops engaged in this vicinity; the 5th New York was involved here also. After the regiment had pushed the Confederates out of town and fought southwest of Hanover, the New Yorkers were ordered to support Elder's battery. Before long, however, their role was changed again. Major Hammond's troops were sent east of Hanover, possibly to the left of the 1st Vermont. At some point, they were ordered to "flank the enemy's position, and capture the battery, if possible...."[256] Although the 5th New York pushed forward they were not able to strike the Southern forces at that point. Apparently, Hampton's men realized the intention of the Union flanking movement and fell back before any large assault took place. Before long, the Union troops were ordered back to assume a defensive posture once more. The 5th New York fulfilled a number of roles that day.

ACTION ON THE CONFEDERATE LEFT

Meanwhile to the west of town, an important situation was shaping up for the Union high command. The status of the Littlestown Road had become an increasing concern to Kilpatrick. This thoroughfare was not only his line of communication with Slocum's Twelfth Army Corps at Littlestown, but also with Gen. George Meade's Army Headquarters at Taneytown, Maryland. Aggressive action by dismounted Confederate skirmishers posed a constant threat to any dispatches sent from Hanover. By early afternoon, it had become critical for Kilpatrick to make sure the road was open for Union couriers. This situation brought about another phase of large-scale skirmishing.

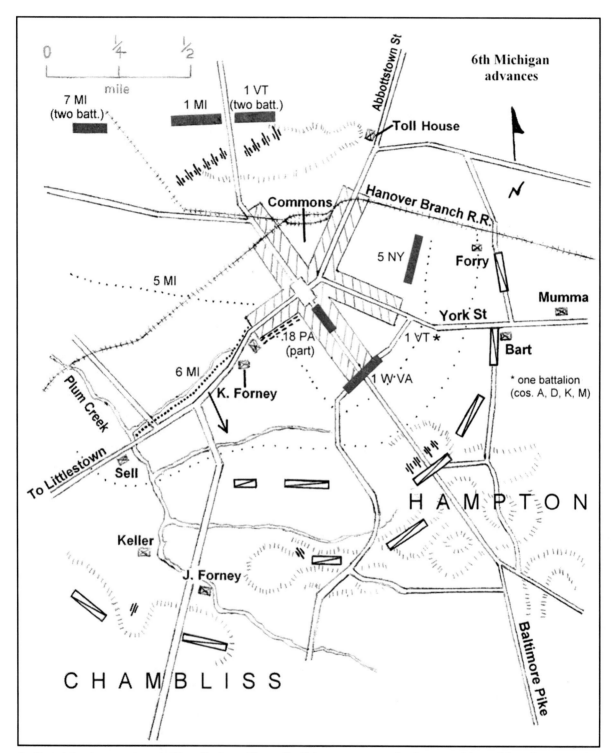

1- Remaining troops arrive. Artillery fire intensifies as Confederate guns open fire from near Mount Olivet Cemetery. 2- After the Schwartz Schoolhouse fight, Lee's brigade begins to escort the wagon train eastward. 3- 6th MI makes dismounted advance. This movement reopens Kilpatrick's line of communications on the road to Littlestown. 4- Skirmishing intensifies along axis of the York Road between one battalion of the 1st VT and portions of Hampton's brigade. 5th NY is ordered east of town. 5- Skirmishing continues until Chambliss and Hampton withdraw.

The 6th Michigan reached Hanover not long after their fight with the 1st Virginia. When Colonel Gray reported to General Custer, the regiment was ordered to dismount and form a line of battle. The 5th Michigan arrived shortly after; they also were deployed on foot.[257]

The staging area for the 6th Michigan's ensuing movements was likely the open ground north of the St. Matthew's Church and west of Carlisle Street. The regiment then wheeled to the south and marched toward the Hanover-Littlestown Road and the Confederate positions beyond.[258] Since one of every four men was tending to the horses, and with companies B and F still trapped behind the Confederate lines, the regiment possibly had fewer than 400 troops available for a dismounted advance. Likely there were several yards between each soldier in the line.

About this time, some of the 6th Michigan became aware of the presence and charisma of their new brigade commander. Captain Kidd (of Company E) stated that as the men were "deploying forward across the railroad into a wheat field beyond, I heard a voice new to me, directly in rear of the portion of the line where I was, giving directions for the movement, in clear, resonant tones, and in a calm, confident manner, at once resolute and reassuring. Looking back to see whence it came, my eyes were instantly riveted upon a figure only a few feet distant, whose appearance amazed if it did not for the moment amuse me."[259] Apparently, George Armstrong Custer's appearance was as striking as his leadership abilities. Kidd's description of the general included the following:

> Looking at him closely, this is what I saw: An officer superbly mounted who sat his charger as I if to the manor born. Tall, lithe, active, muscular, straight as an Indian and as quick in his movements, he had the fair complexion of a school girl. He was clad in a suit of black velvet, elaborately trimmed with gold lace, which ran down the outer seams of his trousers, and almost covered the sleeves of his cavalry jacket. The wide collar of a blue navy shirt was turned down over the collar of his velvet jacket, and a necktie of brilliant crimson was tied in a graceful knot at the throat, the long ends falling carelessly in front. The double rows of buttons on his breast were arranged in groups of twos, indicating the rank of brigadier general. A soft, black hat with wide brim adorned with a gilt cord, and rosette encircling a silver star, was worn turned down on one side giving him a rakish air. His golden hair fell in graceful luxuriance nearly or quite to his shoulders, and his upper lip was garnished with a blonde mustache. A sword and belt, gilt spurs and top boots completed his unique outfit."[260]

It became quickly apparent to Kidd that the general was not just relaying orders, but was in command of the movement. Under Custer's direction, the 6th Michigan pressed on toward the Confederate skirmish lines. After crossing the Hanover-Littlestown Road, their approach was gradually uphill through Karle Forney's land; a low ridge which ran parallel with the road, masked the movement of some of the regiment for a short time. Their line was close to three-fourth mile in length at this point; Henry Sell watched from his house and stated that the advance stretched from the Forney farm to the Plum Creek.[261]

Several Hanover area residents suffered large economic losses that day. Confiscated horses, destroyed fences, and trampled crops were the major damages sustained. Karle Forney stated that Union cavalry marched through his wheat field, trampling and destroying several acres. He also said that "many dead animals were dragged over it after the skirmish."[262]

After reaching the crest of the "Forney" ridge, the 6th Michigan began to exchange fire with the skirmishers of Chambliss. Four companies of the regiment carried the Spencer rifle, while others were equipped with Burnside carbines. With its greater rate of fire and longer range, the Spencer gave those companies a decided advantage. At Hanover, the psychological value may have been important also. This was the first time Stuart's men faced Union troops that carried these superior weapons. It is not hard to imagine the shock of the Southern cavalrymen when they encountered an enemy who could maintain a rate of fire that they themselves could not begin to match.

The appearance of the 6th Michigan also drew artillery fire. Their advance was a threat the Confederates could not ignore. From Stuart's perspective, it may have been the start of a much bigger Union offensive. But the Michigan men did not continue their movement against the higher ground where the main Confederate force was located. Not only was that terrain a dominant factor, but their advance had achieved Kilpatrick's goal of opening the line of communications along the Hanover-Littlestown Road. (It is likely that Lee's Brigade had begun to escort the wagon train previous to this dismounted advance. See following chapter. The fact that the 6th Michigan was able to march a line of skirmishers south of the Hanover-Littlestown Road without instigating a major counterattack suggests that Lee had already left the area. Had Stuart decided to commit even a portion of Lee's Brigade into this phase of the battle, the 6th Michigan would have had a large Confederate force completely outflanking, and even behind, them. Several writers have attributed the casualties of the 6th Michigan to this particular phase of the battle, not realizing the extent of their combat prior to this time. But when one examines the descriptions and context of the soldiers' accounts, it is clear that the great majority of the regiment's casualties occurred in the mounted fighting before they reached Hanover.)

Both sides claimed success in this stage of the fighting. These dismounted advances regained significant terrain, as Confederate skirmishers fell back for at least a few hundred yards. When Chambliss withdrew somewhat later, it gave the Michigan men the impression that their movement had done much to precipitate the withdrawal. The Confederates also claimed a victory of sorts after this phase; they had seen the march of the 6th Michigan stall before it reached the main Southern positions. At least one Southern artillery officer thought that his cannon were the dominant factor in repulsing an important threat.[263]

But each side's perception of the events was based on a lack of knowledge of the intention of their enemy. Kilpatrick's purpose was not to take the terrain held by the Confederate main line. Even with the firepower of the Spencer rifles, a single regiment could not seriously threaten the higher ground held by artillery and a few thousand cavalry. The advance of the 6th Michigan was intended to secure the communication link towards Littlestown, not to charge the Confederate guns.[264] The Union men also misinterpreted the events. The later Confederate withdrawal was not caused by their dismounted advance; Stuart's decision to move away from Hanover was based on much bigger tactical and strategic concerns.

By early afternoon, Stuart believed that his only option was to disengage and detour completely around Kilpatrick's forces; there was no way he could gain control of the route through Hanover that he had intended to use that morning. The general was likely aware by that time that the Union Twelfth Corps had reached Littlestown so any thought of moving to the west was also out of the question. He had to move eastward and hope to completely bypass the Union troops in that direction.

At that time the Confederates still held the high ground south of Hanover. Their "line" stretched in an arc of about two miles with close to 5,000 men (at least while Lee's Brigade was still on the field). Behind their position, the captured wagons were placed in park, near the Westminster Road. It is from these general areas that the eventual withdrawal occurred. Hampton's deployment allowed Stuart to feel much more secure on his right and created conditions that were favorable for withdrawing the wagons, while still keeping a blocking force at Hanover for a time.

Shortly after Lee's arrival, his troops were ordered to take control of the captured wagons for the movement away from Hanover. Lee's men rode behind the higher ground where the other brigades were positioned and began to escort the wagon train. This movement initiated the march that eventually took the Confederates east through Jefferson and close to Seven Valley, where they turned north en route to Dover and Carlisle.

This detour points out the real importance of June 30, which was not so much the fighting itself but the fact that Kilpatrick had reached the critical intersection of the Hanover-Littlestown and Westminster roads before Stuart. Once again, Stuart's attempt to communicate with Confederate infantry had suffered a setback. From the time he had separated from the main body of the army in Virginia, his movements had involved a series of critical decisions, each having an impact on succeeding events. Now the accumulated impact of those decisions was growing larger with each passing hour.

CHAPTER ELEVEN
WAGON TRAIN WITHDRAWAL

Once the decision was made to disengage, other logistical concerns became important for Stuart. Certainly the safety of his men was of paramount importance during the movement. But the captured wagons, along with the Confederates' own vehicles, presented other difficulties. A retrograde movement of three cavalry brigades was one thing, but a withdrawal of well over one hundred wagons in the face of the enemy was another. The road system south of town, which had already been critical that day, then assumed an importance of a different nature. Several roads that led away from Hanover were used by the different elements of Stuart's force when they disengaged. Most of these withdrawal routes were southbound; they, in turn, led to the critical eastbound roads, which were used to move toward Jefferson.[265]

A few Southern accounts stated that the captured wagons were placed in park for a time, but it is unlikely they were in this position for long, possibly even less than an hour.[266] By noon, the tactical situation at Hanover had made an eastward movement almost inevitable. To add to Stuart's concerns, the movements of the 5th and 6th Michigan had given him the impression that the wagons were being threatened.

Large tracts of land south of Hanover were available where the wagons could be guarded; the precise location where they were placed in park is difficult to establish. (See Appendix K for a more detailed discussion concerning this location.) Certainly they would have been placed near the Westminster Road and likely not far from Conewago Hill. Whether the wagons began moving from Conewago Hill or from other farmland near there, the Fairview Road became the initial primary route away from the Hanover area. (A short stretch of Beck Mill Road may also have been used to reach Fairview Road.)

The Joseph Arnold farm, along Fairview Road, received major damage as the column moved from the apex of Conewago Hill toward the Baltimore Pike. Arnold filed a damage claim for three mares taken and three acres of wheat destroyed in the ground. His daughter Sarah also gave a statement, saying, "A large number of rebel soldiers" passed over the premises.[267] Meanwhile, Ephraim Nace, after he was taken by one of the Confederate details, watched as "a long train of army wagons, one hundred or more, came across West Manheim Township to the Baltimore Turnpike."[268] Although on higher ground than the town of Hanover itself, Fairview Road follows a relatively flat course and was probably one of the least demanding sections of the Confederate movement that day.

Upon reaching the Baltimore Pike, the column traveled along that road for a short time before turning east onto Fuhrman Mill Road. Nace continued to observe the movement and stated that the train moved along the Pike "to what is now known as Centre school-house, above the Brockley farm, and passed toward Jefferson." He also met Stuart personally, whom he described as "a middling tall man with a full reddish brown beard." Nace asked the general to "let me go but he would not consent, stating that he needed us."[269]

Near the Baltimore Pike/Fuhrman Mill Road intersection there occurred an incident, which led to several shots fired. A Confederate officer inquired of a farmer whether any Union soldiers were in the area. The local man knew some Union soldiers were concealed in woods a few hundred yards away, but as he wanted to avoid an incident, he denied any knowledge of their presence. At this point, his young son, apparently raised with a strong commitment to honesty, stepped up and corrected his father and proceeded to point out the area where the "squad" of Union men was concealed. Nace witnessed this event and said a "Company of Confederates charged into the woods after the Federal soldiers and several shots were exchanged. Some horses were shot and I think one man was wounded or killed. When they began to shoot, an officer turned around to us and shouted, 'Get off your horses, you Dutch Yankees, or you will be shot', and all who were acting as guides jumped down from our horses and hugged the ground until the shooting was over."[270]

If the teamsters driving the wagons thought they had seen the last of challenging terrain, they probably would have done some excessive swearing when they made their way along Fuhrman Mill Road. Similar in nature to the demanding hills on the Hanover-Westminster Road, this stretch of the movement also created similar hardships. As the column moved eastward through West Manheim Township, several landowners were affected. Three mares and one horse owned by William W. Allbright were seized as the Confederates passed his residence. Henry Rudisill witnessed this incident and said that while he stood there with the rebel army passing him along the road, he saw some soldiers enter Allbright's field and gather the animals.[271] Simon Barnhart, who also resided along Fuhrman Mill Road, had forty bushels of corn confiscated, along with two mules, a buggy, and a harness.[272]

The column continued to the end of Fuhrman Mill Road. At this point, they likely crossed Black Rock Road, then used the lane that passed the John Wildasin farm to reach Dubs Church Road.[273] Well before this time, word of the Confederate presence had already spread through the area. Several citizens tried to hide their animals or move them to safety. At least three men who lived nearby took horses to a more secluded area. The animals were hidden among brushy cover in a meadow owned by George Millheim but were later discovered by a detachment of four or five Confederates in the bushes east of Dubs Church Road. According to one witness, it was "at about noon or a little after that time" when the Southern soldiers found these animals.[274]

The cavalrymen seemed to be getting much more effective at finding the horses than the civilians were at hiding them. One local historian has estimated that the Confederates gathered 265 horses from West Manheim, Manheim, and Codorus Townships alone; others believe the total to have been higher. Probably at least 1,000 animals were seized from York County overall by the time the campaign had ended.[275]

(The following stage of the Confederate movement, which took them from Dubs Church Road to the Sinsheim [Jefferson] Road, is challenging to reconstruct. With the addition of Codorus State Park (more than 3,300 acres), including the 1,275-acre Lake Marburg, some significant road changes have taken place, and much of this section of Stuart's ride is literally underwater. To accurately trace the movement through this area requires not only historical maps but also a modern fishing map that shows the submerged roadbeds.)[276]

The column continued to the Dubs (St. Paul's) Church, then moved somewhat eastward along a road from the church into the lower ground now covered by Lake Marburg.[277] The area around that house of worship saw a tremendous amount of Confederate activity that day; at least three routes taken by Stuart's men converged near there. Almost every soldier in his force would have traveled within sight of this landmark as they made their way toward Jefferson.

William Dubs operated a mill in this area and probably remembered that day for a long time. Dubs listed the following as some of the items taken: 34 bushels of wheat, 40 bushels of rye, 315 bushels of corn, and 240 bushels of oats.[278] A neighbor, John Snyder, lost 16 bushels of corn, 23 bushels of oats, and 24 bushels of rye.[279]

Further east, some Confederates encountered resistance from Luisa Dubs. Luisa, at home alone, pleaded with soldiers not to take her horses, at which point an officer ordered the men to put the

A partial listing of either state or federal damage claims for items or animals taken. Numbers on the map indicating locations where goods were confiscated. Evidence from the claims seems to indicate that, with the exception of #10 and #11, locations west of the Baltimore Pike were visited by Confederate details before or during the fighting at Hanover. Those east of the Baltimore Pike occurred while Stuart's men moved away from Hanover. Although claimants often distinguished between mares, stallions, and horses, for simplicity I have listed all as horses:

1-Josiah Gitt-3 horses, 3 mules, 75 bushels of corn, 20 bushels of oats, 30 grain bags (Gitt resided in Hanover; these items were taken from his farm along the Westminster Road which was worked by Edmund Lippy.)
2-Perry Mathias-2 mules, 2 collars, 2 bridles
3-David Baughman-3 horses
4-Benjamin Wentz-1 horse
5-Jesse Houck-2 horses
6-Ephraim Nace-4 horses, 4 halters, 1 saddle, 1 whip
7-Andrew Sterner-2 horses
8-John Lohr-2 horses
9-Jacob Leppo-3 horses
10-Edward Becker-1 horse
11-Joseph Arnold-3 horses, 3 acres of growing wheat destroyed
12-Franklin Emmert-1 horse
13-Anthony Brockley-2 horses - 1 other seized from his son, John
14-Jacob Shue-2 horses
15-Simon Barnhart-2 mules, one buggy and harness, 40 bushels of corn
16-Jacob Mummert-2 horses
17-William Albright-4 horses
18-Jacob Wise-1 horse
19-John M. Wildasin (Heidelberg Township) -2 horses (another horse seized on June 27 by 35th VA Battalion)
20-John Wildasin (West Manheim Township)-1 horse
21-Jacob Meyers-1 horse
22-Jeremiah Millheim-1 horse
23-Christian Millheim, Levi Christman-3 horses seized from Christman; 2 from Millheim. 4 of the 5 of were hidden on land of George Millheim in this vicinity.
24-George Baumgardner-4 horses
25-Michael Runkle-1 horse - another family member, David, had 2 horses seized
26-Catherine Wildasin-3 horses
27-John H. Snyder (Shuyler)-16 bushels corn, 23 bushels oats, 24 bushels rye, 16 empty bags (plus one horse taken on June 27 by 35th VA Battalion)
28-William Dubs-1 horse, 1 set of hind gears, 1 saddle, 1 wagon line, 34 bushels of wheat, 40 bushels of rye, 315 bushels of corn, 240 bushels of oats (another horse seized on June 27 by 35th VA Battalion)
29-John W. Snyder- (taken on June 27 by 35th VA Battalion - 1 horse, 1 saddle, 1 bridle)
30-George Dubs-Dubs filed no claim. Apparently, when Confederates found two horses in his stable, his wife Luisa protested enough that an officer ordered the men to leave the animals.
31-Daniel Runkle-1 horse
32-John Miller-1 horse
33-John Kitzmiller-1 horse
34-Ephraim Leppo-1 horse
35-George Pressel-1 horse
36-Jacob Miller-2 horses (another horse seized on June 27 by 35th VA Battalion)

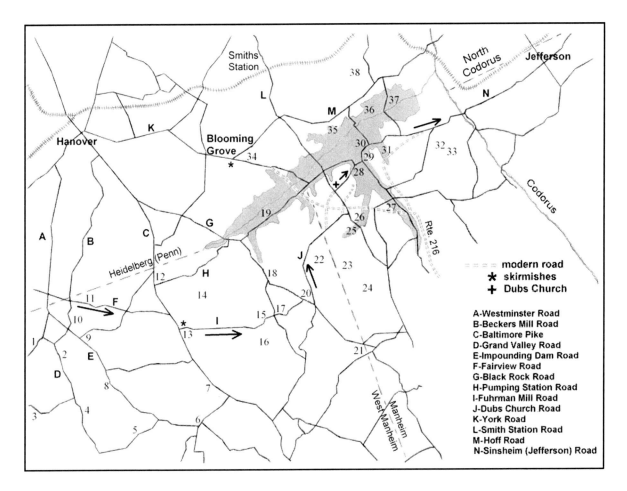

STUART WITHDRAWS FROM HANOVER SOUTHWESTERN YORK COUNTY

Darker roads are Civil War era traces. Road names are mostly post-war. Lake Marburg and selected modern roads are superimposed to show changes in the Codorus State Park area.

1- Lee's brigade begins to escort the captured wagons. The column moves along Fairview Drive, turns south onto the Baltimore Pike, then eastward on Fuhrman Mill Road to Dubs Church Road.
2- The column likely travels the (original) length of the Dubs Church Road to its intersection with the road to Smith Station. They ride northward for a short time and turn right at the Dubs Church. (See endnotes 277, 279 for Chapter 11.)
3- The Confederates move past Dubs Mill and into the area now covered by Lake Marburg. Continuing eastward, they ride through Jefferson and toward Seven Valleys. Before reaching Seven Valleys, they turn northward and head toward New Salem and Dover.
4- Remaining Confederate forces begin to withdraw from Hanover area. Chambliss, for the most part, follows Lee's route. Hampton, as the rear guard, uses the Baltimore Pike, Black Rock Road, and the York and Blooming Grove Roads. At least two skirmishes occur during the withdrawal; one near the Baltimore Pike/Fuhrman Mill Road intersection (the "Brockley Farm Affair"), another in the vicinity of Blooming Grove (post-war name). Asterisks on map indicate general areas of skirmishing, not precise locations.
5- As Stuart moves, his troops continue to seize goods, particularly desperately needed horses. Detachments use many connecting roads and farm lanes to scour the surrounding countryside.

1900 view of Dubs Church looking northward. (GPL)
The lower ground behind the church is now covered by Lake Marburg. Hoff Road runs along the ridge in the distance. Stuart's main body rode past the church as they headed toward Jefferson.

horses back in the stable.[280] Traveling east, the column moved out of the area, now covered by the lake, and continued toward Jefferson.[281]

To the north of the main column, small detachments crossed the west branch of the Codorus Creek to search for more horses. One local man, Henry Rennol, stated that he left his home "about 3:00 to go to a store about a mile distant." Upon hearing the Confederates were moving through the area, he returned home "being gone only about one half hour," and discovered that his two horses had been seized. Apparently, when Rennol learned that his family had seen only one or two Confederates take the animals, he got the notion he could get them back and started to follow the course the soldiers had taken. He was disabused of this notion shortly, as after going "about one-fourth mile," he discovered "a large body of rebels." Rennol likely had sighted the main body, and trying to retrieve his horses suddenly did not seem like such a good idea. He admitted that at this point, he turned around and rode home.[282] With the timing stated, Rennol's animals were likely taken by details from Lee's Brigade, although other Confederates also moved through that same general area later that day.

CHAPTER TWELVE
CHAMBLISS, HAMPTON WITHDRAW

A sense of uneasy stability hung over the battlefield by late afternoon. Although a tactical standoff had been in effect for a few hours, another attack was still possible, and each general was uncertain as to his opponent's next move. Much of Farnsworth's brigade held the town itself, although several companies were still deployed as skirmishers on Hanover's perimeter. Custer's brigade, for the most part, was positioned either on Bunker Hill or on farmland along the axis of the Hanover-Littlestown Road. Sporadic shooting continued as the skirmish lines jockeyed for position. Meanwhile, occasional artillery rounds screamed through the air from both sides of town.

Once the wagon train was safely away from Hanover, and with no apparent Union pursuit being mounted, it became safe for Chambliss and Hampton to pull back from their holding positions. Accounts indicate that some Confederates remained in the Hanover vicinity until at least dusk before leaving. Lt. George W. Beale, Company C, 9th Virginia, stated, "After fighting the enemy for several hours with our sharpshooting and shelling the town quite furiously, thereby giving most of our men time to get around the town and get several miles away, we withdrew without being pursued."[283]

While most, if not all, of Chambliss's men followed the main column route eastward from the Baltimore Pike, how they reached the Baltimore Pike/Fuhrman Mill Road intersection is not certain. Likely much of this brigade retraced its original path on the Westminster Road, ascended Conewago Hill, and then turned left unto the Fairview Road, following the main column route from that point on. This route would have allowed them to guard against any pursuit toward the rear of the wagon train while Hampton was still in a holding position. Other elements of Chambliss's Brigade may have taken the Beck Mill Road south to Fairview Road.[284]

While Hampton remained in position south and east of Hanover the Confederate moving column was protected from any threat to its left flank; Hampton's Brigade was essentially a static blocking force at that point. Once his men began to withdraw, their role changed; they then became a moving, screening force. After leaving the battlefield, they also became the rear guard of Stuart's forces. Most of the brigade likely joined the path of the main body by the most direct route; southward along the Baltimore Pike and Black Rock Road.

At least a few cavalrymen turned east on Pumping Station Road. John M. Wildasin owned a farm along this road; Southern soldiers visited his property at least twice that day. Sarah Huggins watched as a "squad" of Confederates took a bay horse from Wildasin's fields "towards evening."[285] This road eventually merged with the main route near the Dubs Church, bringing these detachments back with the main force. (Much of this area is now under Lake Marburg.)

Another path of withdrawal for some of Hampton's men was the York Road. This route had been critical for Confederates to control during the engagement to prevent the possibility of a Union flanking movement. By late afternoon, it continued to be important; Hampton needed to guard the road to protect Stuart's moving column from any potential Union pursuit from the east of Hanover. It is worth noting that Stuart had assigned the task of screening his movements that day to his two most experienced commanders, first Fitzhugh Lee, and then Hampton.

After moving out the York Road, those elements of Hampton's Brigade then turned onto the Blooming Grove Road. Along this route and about two to three miles east of the Hanover square was a small cluster of houses, which is today known as Blooming Grove. A few shots erupted near the village as troops reached the area. According to Elizabeth Stover, while she was at the Ephraim Leppo residence, "a skirmish took place in the neighborhood."[286] Leppo also referred to shooting that occurred nearby. Apparently a Union patrol had followed some of Hampton's men this far to the east of the main battlefield. At that point a brief, and likely very small, fight broke out. According to Cpl. John Jackson of the 5th New York, after the Confederates withdrew from Hanover, "...a detachment from our regiment under Major Hammond followed them four miles, skirmishing with their rear guard."[287]

Most of Hampton's men who rode through this area likely continued south and intersected the path of the main body near the Dubs Church. But at least a few detachments were active along Hoff Road, which branches off from Blooming Grove to the east. The terrain in the vicinity offered some advantages for the Confederate movements. Much of Hoff Road follows higher ground which parallels the main column route south of there. Southern patrols had the security of being able to see their own brigades to the southeast, while they scouted, gathered more horses, and provided an early warning system against a Union pursuit. At least ten citizens who lived on or near Hoff Road witnessed small patrols seize horses from their property. Some saw only one or two soldiers; others saw details of four or five.[288] Since all three Confederate brigades passed by within one mile of these properties, it is not certain whether they were seeing detachments of Lee, Chambliss, or Hampton.

ROUND ISLAND (AC)
View of Lake Marburg looking northward from the modern Sinsheim Road. The lake covers a portion of Stuart's ride. His main body moved from the left to the right, passed the "Round Island," and continued toward Jefferson. Hoff Road runs along the higher ground in the distance. Confederate details moved through that area.

John Miller, who lived south of the path of the main column, was one of the final Manheim Township residents to be visited that day when three Confederates took a horse from his stable. This farmer spoke to the soldier who had possession of the animal and told the man not to take the horse because she was too young. Apparently this plea did not impress the Southern cavalryman; Miller said that the "rebel was not disposed to comply" with his request, and "spoke saucy"" to him.[289]

It would be safe to say that many Confederates were feeling "saucy" at this point. On June 30, they had already traveled from Carroll County, Maryland, through the southeastern portion of Adams County, Pennsylvania, then into York County and the fight at Hanover. As they continued eastward, the possibility of Union pursuit lessened with each passing mile, but their ride was far from over.

Not long after leaving Hanover, the great majority of Chambliss's and Hampton's men had begun to follow the path of Lee's Brigade. By then, however, there was a several mile gap between the front and rear of the column. The leading troops had reached the village of Jefferson by late afternoon; the other brigades would go through that area later.[290]

Stuart's chief ordnance officer, Capt. John Esten Cooke, reached Jefferson "at dark," where he met a "pretty Dutch girl, who seemed not at all hostile to the gray people, and who willingly prepared me an excellent supper of hot bread, milk, coffee, and eggs fried temptingly with bacon." The officer never got to enjoy the meal, however. About that time, Stuart sent Cooke on an errand, and when he returned, he found himself to be "the victim of a cruel misfortune." His food was gone. The young girl had placed the supper "on a table in a small apartment, in which a side door opened on the street; through this some felonious personage had entered - hot bread, milk, coffee, eggs, and ham, had vanished down some hungry cavalryman's throat."[291]

Jefferson, like Hanover, had experienced the arrival of Lt. Col. Elijah White's 35th Virginia Battalion a few days before. Now on June 30, the common theme was still horses; hundreds more were gathered as Stuart's troops moved through a portion of Codorus Township, then cut a wide swath through North Codorus Township.

THE COLUMN TURNS NORTHWARD

After moving through Jefferson, the column continued eastward in the direction of Seven Valleys. Before reaching that village, however, most of Stuart's forces turned onto Panther Hill Road and began the northward trek toward New Salem and Dover. This particular stretch of the movement saw some rather interesting encounters. A funeral service was in progress at Zeigler's Church as Fitzhugh Lee's Brigade approached. Apparently, the burial service came to a quick halt as the cavalry was observed moving toward the cemetery, and many locals rushed home to try to hide their own animals.[292]

Surprisingly, some Adams County men had an encounter with Confederates this far east into York County. Peter Diehl operated a tannery in New Oxford and a wagon freight service between there and Baltimore. With the nature of his business in hauling produce and other merchandise, horses were of vital importance. By June 30 Diehl believed the Confederate threat was too great to ignore and asked his son Charles to move the horses across the Susquehanna River. Accompanying Charles were also Henry and Jacob Diehl, along with David Peters, Elias Slagle, and Amos Lough. These New Oxford area residents had moved the animals several miles, but their luck ran out when they reached the area near Seven Valleys. As they made their way "along the public road leading to Zeigler's Church," they were startled to see Confederates with a portion of the supply train bivouacked in a grove by a spring. At this point, the Adams County men were forced to carry water and help dress the wounds of some of the soldiers. Toward evening, the Confederates resumed their march, and the men were released and given several worn-out horses to return home.[293]

Meanwhile, the Henry Hoff family, who resided a few hundred yards off Panther Hill Road, watched in fear as Stuart's forces rode by. When the end of the column finally passed, the family thought their problems were over until they saw eight cavalrymen leave the main body and ride straight to their house. These soldiers took food and clothing from the house and then horses from the barn. Mrs. Rosanna Hoff was greatly disturbed that one of the horses taken was her personal favorite. The Confederates also discovered Henry's whiskey supply, which apparently was ample as he operated a distillery. Although they loaded as much whiskey as they could carry, they had only ridden a short distance before deciding that a better course of action would be to lighten their load by drinking some of the haul. As dusk fell and the family continued to watch, it became apparent that the more the soldiers drank, the less inclined they were to move, until finally Rosanna noticed they were resting and had not moved for some time. Determined to get her favorite horse back, she quietly made her way down the road to where the soldiers were asleep in some woods and proceeded to retrieve her horse while they slept. The soldiers later awakened during the night and either did not notice the missing animal, or did not care, and rode on to rejoin their column.[294]

Hunger was starting to become a factor for Stuart's men. Sgt. Robert Hudgins II described an event that occurred while on a foraging expedition with two of his comrades. Hudgins, Joe Ham, and "Bang" Phillips knocked on the door of a farmhouse, which was answered by a woman carrying a baby of "about six months." When the soldiers asked to buy some food, she responded, "I have nothing and if I did, I'd not give it to you." Phillips edged closer to the woman and asked again, at which point she stated, "You dirty Rebels will get nothing from me. I'd like to see the whole murderous lot of you die." According to Hudgins, at that point "Without warning 'Bang' snatched the baby from her arms. She let out a scream and it was obvious all her former bravado was gone. 'Bang' threatened her with the possibility, although untrue, that we'd have to eat the baby if we couldn't get something else. He then offered to trade her the baby for some bacon, whereupon she set out ham, fowl, bacon, bread, and butter. We had a glorious feast and took the remainder back to camp after paying her (in Confederate money) for all we had taken."[295]

At some time that night, Stuart met with some of his officers in the home of local resident John E. Ziegler. Ziegler was probably the wealthiest man in the Seven Valleys area, and his farm and mill were almost a magnet for any troops that passed by. Property was seized by not only Stuart's forces, but also by Union cavalry of Gen. David McMurtry Gregg's division that later moved through the area. The losses to Confederates alone included 6 horses, 6 tons of hay, 35 bushels of oats, and a staggering 500 bushels of corn taken from Ziegler's mill.[296]

At this point, Stuart was safely beyond Kilpatrick's reach. However, there was still no indication that he was close to establishing communications with any Southern infantry. It is problematic to theorize what Stuart's state of mind was at any given time, but certainly a sense of urgency would have guided his movements before Hanover; after the battle, it appears that there was almost a desperate quality to the march. The men badly needed rest, but it could not be given. The column had to push forward, cover as many miles as possible, and search for signs of Confederate infantry. The march went on into the darkness and became an endurance test, which strained men and animals beyond the breaking point. The accumulated wear and tear of the previous days began to induce a sleep-deprived stupor in many of the men. Some fell asleep in the saddle, and occasionally a soldier would fall off his horse onto the road. Lt. George Beale stated, "From our great exertion, constant mental excitement, want of sleep and food, the men were overcome and so tired and stupid as almost to be ignorant of what was taking place around them. Couriers in attempting to give orders to officers would be compelled to give them a shake and call before they could make them understand."[297] Although this statement described the condition of the men at Carlisle on the following evening, it could just as easily have applied to the night march of June 30.

After failing to procure a meal in Jefferson, Capt. John Esten Cooke had better luck in New Salem. A few officers stopped in a local tavern where they were able to get some food and drink. Stuart drank a cup of coffee then went ahead with the column. In the meantime, Stuart ordered Cooke to remain in the village until Hampton arrived, and then direct him to Dover. The captain had to wait a while; the brigade was "several miles behind" at that point. Cooke was overcome by exhaustion and fell asleep in a chair on a porch, when a "clatter of hoofs resounded, and the voice of General Hampton was heard in the darkness..."[298]

Stuart's frustration continued the following day. His forces crossed over the York Road a few miles west of that city, and the head of the column reached Dover in the early morning of July 1. By then he learned that General Early's Division had pulled back from York and retraced their movements westward. Delays, detours, engagements, and decisions made by Stuart had repeatedly pushed his troops farther away from Southern infantry. Ironically, at the very time when the cavalry was on the verge of communicating with the infantry, Gen. Robert E. Lee began to gather his army toward the Gettysburg/Cashtown area, pulling farther away from Stuart.

When Stuart's forces reached Dover, at least some of the civilians who had been forced to accompany their movements as guides were still with them. But, by that time, the usefulness of these men was doubtful since the cavalry was now moving in an area unlikely to be familiar to any Hanover area residents. Ephraim Nace was placed in a house under guard while several Southern officers breakfasted at a Dover hotel. He escaped down a back alley and climbed into a haymow, which had been filled with fresh hay. He remained hidden until almost all of the Confederates had left Dover. Apparently, his presence was not considered critical anymore. Nace related that there were still a few Confederates in the town, and they gave him "two old worn out nags that had been in the Confederate service for a long time. I rode one of these horses home and led the other. One of the horses died within a week and the other I was glad to sell for $10. What became of my faithful horses, I know not. Possibly they were killed in the battle of Gettysburg."[299]

Clearly, the separation of Confederate infantry and cavalry had far-reaching and critical effects on the course of the campaign and the war, effects which are still debated to this day. For many citizens like Ephraim Nace, the impact was immediate and much more apparent.

CHAPTER THIRTEEN
THE NEXT CRITICAL HOURS

By the evening of June 30, Union troops had controlled Hanover for several hours. Stuart's attempt to move through the town had been thwarted, and Kilpatrick's communications with army headquarters had been reestablished. Most of the Union regiments had made no major movements since mid-afternoon, although some shifting and consolidation of positions did occur.[300]

Meanwhile, the officers of the various units examined casualty reports and assessed the condition of their units. Of the Union regiments, the status of the 18th Pennsylvania was the greatest question mark. Several companies of the 18th had been decimated and scattered; as the afternoon and evening wore on, various Pennsylvanians rejoined the regiment.

While the battle readiness and marching condition of the division was being mended, intelligence was another major concern. Stuart had disappeared from Kilpatrick's front, but his destination was unknown. The Union general began to send patrols to gather information and it soon became apparent that the Confederate forces had moved well away from Hanover.

Kilpatrick possibly would have liked to pursue the Southern column aggressively, but if so, that action was prevented by much bigger strategic concerns. The major role of his division since leaving Frederick had been to screen Union infantry and gather intelligence on the marches of Confederate infantry, not to carry out an unchecked pursuit of Stuart. It was critical for his movements to stay within the parameters assigned to him by army headquarters. An unintended battle at Hanover did not give Kilpatrick the latitude to take his division out of communication and supporting distance of the Army of the Potomac. By this time, Gen. Robert E. Lee's infantry was moving blindly without an effective cavalry screen; General Meade could not afford to let that lapse in protection and communication happen to the Union troops. Had Kilpatrick chased Stuart through York County, an important "piece" would have been taken "off the chessboard," and a large hole would have been left in the Union cavalry screen at a very dangerous time in the campaign.

Much of the fighting that day had involved control of the Hanover-Littlestown Road. Since the Twelfth Corps was encamped at Littlestown on the evening of June 30, that route remained a vital artery of communications. Kilpatrick's division bivouacked that night in the Hanover area, mostly on the outskirts of town.[301] Some campsites were north of town in the direction of Abbottstown. But, in almost every case, these units were situated on properties that bordered their line of communication.

According to Kilpatrick's official report, casualties for his division on June 30 totaled 197. (Note: Ironically, the numbers that Kilpatrick gave for killed, wounded, and missing actually add up to 183.) Southern losses, although more problematic to determine, were at least 160.[302]

At about 2:00 P.M. of July 1, while fighting raged near Gettysburg, the Third Cavalry Division rode toward Abbottstown, essentially traveling along the axis of the Adams-York County line. Kilpatrick's men reached (East) Berlin and beyond; scouting details probed the roads to the north and east, with one detachment riding as far as Rossvile. Meanwhile, Stuart's troops, still on the perimeter of the operational area, moved northward through central York County, with a few thousand of the enemy between themselves and their intended objective.

These counties were bountiful and rich with crops; soldiers on both sides occasionally used the term "magnificent" in their descriptions of the countryside. Both sides continued to seize supplies, especially from local farms and mills. Significant numbers of Custer's brigade encamped for a time on properties near the York Road. Just east of Abbottstown, Peter Noel had recently harvested some hay, which was then piled in his fields. His timing could not have been worse, for a large number of the 6th Michigan carried off three to four tons of the cut hay. Other soldiers seized about twelve bushels of shelled corn from his barn. Noel did receive a receipt for his losses; Lt. D. G. Lovell, 6th Michigan, who was serving as the regimental quartermaster at that time, signed this voucher.[303]

One neighbor of Noel's was not so fortunate to receive a receipt. A part of Custer's command encamped on the John Myers farm for "part of two days," in which time he lost several bushels of corn and oats, along with almost one ton of hay. Myers' provisions provided not only for the horses, but also for the troops themselves; they seized 400 pounds of pork, while he personally used two barrels of flour to bake for the soldiers.[304]

The day before, Rev. John B. Catani had ridden from the Conewago Church to minister to wounded soldiers in Hanover. After July 1, he would find his vocation much more difficult to fulfill. The priest's horse was seized by Union cavalrymen from its stable near the Paradise Church. (Located near Abbottstown, this structure was a mission of the Conewago Church.)[305]

Around midnight, most of Kilpatrick's men had retraced their movement back to (East) Berlin and bivouacked near that town. The next day, the general learned of the fighting at Gettysburg and was ordered to report to that area as quickly as possible. His troops retraced their route from (East) Berlin to Abbottstown and then moved westward along the York Road. As the Third Division approached Gettysburg, they made contact with Union troops of General Gregg's Second Cavalry Division along the York Pike. From this location, Kilpatrick and Gregg could guard against any possible flanking movement toward the Union right. Soldiers of both divisions would soon encounter some of the same Confederate cavalry that Kilpatrick had met at Hanover.

At Dover on the morning of July 1, Stuart was faced with another important decision. He had been "informed by citizens" that Early "was going to Shippensburg."[306] It seems doubtful that Stuart would have put much faith in the specific destination; the information, after all, was given by civilians in a hostile area. But the most important aspect of the intelligence was easy to verify; Early had moved westward from York. Clearly the Confederate movement toward the Susquehanna had been reversed. Since Stuart had failed to connect with the right flank of the Confederate infantry, he decided to move toward where he believed its main body would be. He stated that "I still believed that most of our army was before Harrisburg, and justly regarded a march to Carlisle as the most likely to place me in communication with the main army."[307]

Some of Stuart's forces had been able to take a short rest at Dover before taking up the march once again. The column was still stretched out for several miles at this point, with Lee in front, followed by Chambliss, then Hampton. The vanguard rode through Dillsburg then continued northward.

Confederate frustration continued when they reached Carlisle that afternoon. This town, like York, was one that had been held by Confederate infantry before Stuart reached the area. Once again, his cavalry had just missed a vital connection. To add to his woes, his troops desperately needed food by this time, but Union militia was holding the town. An officer from Lee's Brigade was sent under a flag of truce to demand that the town surrender; the alternative was that Carlisle would be shelled. The terms were offered on two separate occasions and each time they were refused. At about 10:00 P. M., an artillery bombardment began from Confederate guns positioned on hills outside of town. The cannon fire continued intermittently for the next few hours. A third demand for the town's surrender was offered about midnight, but this one was also turned down.

While Confederate forces surrounded the town, a message was received from Gen. Robert E. Lee that the Army of Northern Virginia was engaged at Gettysburg. Stuart sent orders for Hampton to start immediately for that town and also issued marching instructions to the other brigades. He personally began to ride toward the battlefield also. Since Hampton was in the vicinity of Dillsburg at that time, he turned his men onto the Harrisburg-Gettysburg Road and began to move southward, even while the other two brigades were still near Carlisle. As a result, Hampton, who had been last in the march after Hanover, then became the first brigade in line on the approach to Gettysburg.

CHAPTER FOURTEEN
BACK IN HANOVER.....

Soon after the fighting started at Hanover, wounded men made their way, or were carried, into various homes and structures. Before the final shots were fired, a few public buildings, which had served as impromptu medical sites, became officially designated hospitals. These locations included the following ones listed in a Hanover newspaper: "Eckert's Concert Hall, on Market Square, Marion Hall, on Foundry Alley, Albright's Hall, on Broadway, and Pleasant Hill Hotel, on Baltimore Street."[308] Wounded soldiers who had originally been in people's homes were also transferred to these sites.

At least five local physicians attended to soldiers, either in the streets or in the hospitals. These men included Doctors Bange, Hinkle, Culbertson, Eckert, Smith, and Alleman.[309] But the overall responsibility for the hospitals fell upon medical personnel of the Army of the Potomac. On July 1, Dr. Perrin Gardner, assistant surgeon of the 1st West Virginia, was ordered to remain in Hanover and take charge of the care of the wounded. He reported to Marion Hall with a staff of stewards and nurses and quickly began to establish order in a scene of confusion. Before long, carpenters had built bunk beds for the wounded. Food and drink was delivered by several local citizens, and the staff began to organize the medical supplies from Kilpatrick's division. Meanwhile, Gardner went about the task of conducting surgery on the more seriously wounded.

The bodies of the Union men who had been killed were taken to an apartment in the Flickinger Foundry on York Street. Later that evening, they were placed in caskets and buried in the cemetery of the Reformed Church, where the Reverend Zieber performed the last rites. Meanwhile, a few other Northern soldiers were interred in the graveyard of St. Matthew's. Eventually the Union remains were exhumed; some were reburied in the Gettysburg National Cemetery. On the Confederate side, several of Stuart's troops had fallen in areas that were then controlled by Kilpatrick's forces. Some were buried by civilians on various properties, including at least a few on the Karle Forney farm. In the years after the battle, their remains were exhumed and removed to Southern cemeteries.[310]

On July 9, the wounded were taken to the Pleasant Hill Hotel on Baltimore Street. From that time on, this location was the only significant hospital in the town. This structure was large, with three stories and an attic. *The Hanover Spectator* included the following description of the site: "The building is most admirably adapted for Hospital purposes, being located on high ground, with large yards attached, and commodious verandahs in front of the edifice, where the convalescents can set and have a commanding view of the beautiful and diversified country for miles around. The rooms are also large, roomy, and well ventilated, which is a very great desideratum. The soldiers are all highly pleased with the change."[311]

The hospital staff not only tended to those wounded at Hanover, but also many Gettysburg casualties. Area residents, particularly the Ladies Aid Society, who received the following praise from Dr. Gardner, provided much help: "...the citizens are awake to the highest degree of kindness and generosity in behalf of the soldiers. Every desired comfort is furnished in great abundance and every luxury with which the country abounds in rich profusion is flooded upon by the delicate hands of the good ladies, and in most instances administered to the suffering wounded by the same hands. A more hearty response to the calls of humanity never came from a generous people than we have witnessed here. The Ladies Aid Society is largely attended every day, and large quantities of bed clothing, bandages and every necessary dressing and appliance is here manufactured to supply the present demand."[312]

The care shown to the casualties often had a deep impact on their families. One Michigan man, whose son had been wounded, wrote a letter to a newspaper expressing his gratitude. This father stated that a citizen had picked up and conveyed his son to Hanover, where he received care from the ladies of the town. He went on to say, "These women came to our soldiers like angels of mercy. Truly their mission is a holy one; and Michigan, as well as every other state, owes them a debt of gratitude drawing interest for all time."[313]

Certainly the hospital sites were scenes of much agony and suffering, even with the good intentions and actions of many locals. But another cause for concern soon emerged. In the aftermath of Gettysburg, stragglers began to roam the area, and intoxicated soldiers were creating major disturbances in Hanover. Dr. Gardner described the problem as "gigantic" and said that "the quietude of night was disturbed by the Bacchanalian revels of drunken men." Before long, Gardner asked the borough officials to stop all sales of liquor in the town as long as the hospitals remained open. This ban achieved the intended effect. The doctor later reported that "since the closing of drinking houses and saloons, there has not been one drunken man on the streets or in the alleys, and good order prevails in the hospital."[314]

By mid-July, other difficulties began to occur. Hundreds of people traveled to Pennsylvania in search of loved ones who had been killed or wounded at Gettysburg and related battles. One train that arrived in Hanover on July 9 brought so many people that the town's hotels were filled. Several residents took in many of these strangers to their own homes.[315]

Sometimes these journeys ended happily if a wounded family member was found recovering in a hospital. Often they ended in tragedy. Eber Cady, 18th Pennsylvania, had been wounded during the battle; by mid-July, he was still in Hanover, fighting for his life in a hospital. The Cady family received word of his wound, and by July 21, Eber's sister had arrived at Hanover, after traveling from the family's home in Crawford County, Pennsylvania. Unfortunately, by that time, the soldier had also contracted typhoid. The combination of battle wound and illness was fatal; Cady died on July 26. In a letter to his parents a few months before, Eber wrote that "I still live in hopes... and think that I will be at home after a while, but - Dear friends if we never meet on earth let us meet in heaven."[316]

About one week after the battle, U. S. Army Medical Inspector Lt. Col. Edward Vollum was ordered to Gettysburg for duty concerning the transportation of the wounded. This officer spent a few hours in Hanover, where he found about 150 men under the care of Dr. Gardner. Vollum reported that the wounded were "comfortably situated" and that "the inhabitants had furnished them with bunks, bedding, dressings, utensils, and food in sufficient quantity, the people in each street in the town furnishing food, delicacies, nurses, [for] two days at a time."[317]

Over the next several weeks, many of the Hanover wounded were evacuated to Harrisburg, Philadelphia, and Washington, among other areas. On August 15, the Pleasant Hill Hotel officially closed its doors as a hospital. Throughout the summer and fall, several thousand of the casualties from Gettysburg passed through Hanover by train, on their way to the large Federal hospital sites in the bigger cities.

Certainly the number of casualties and the scale of the suffering at Hanover did not approach that of many other Civil War engagements. Yet Hanoverians could be justifiably proud of the role they had played. Their actions in feeding the troops and helping to alleviate the suffering had won praise throughout military and civilian circles.

EPILOGUE

The Battle of Gettysburg was well under way by the time communication was reestablished between Lee and Stuart. But the participation of the cavalry in the campaign was far from over. On July 2, elements of Hampton's and Custer's brigades made contact at Hunterstown, another engagement in which the consequences far exceeded the number of troops involved. With Hampton involved at Hunterstown, he was unable to screen the Confederate left flank at Gettysburg while Southern infantry attacked Culp's Hill.

On July 3, Kilpatrick's men would see battle in two separate actions. Custer's brigade engaged with Stuart about three miles to the east of Gettysburg, on what is now known as East Cavalry Field. Meanwhile, Farnsworth's men were involved in attacks on what is now called South Cavalry Field, an action that would cost Brig. Gen. Elon Farnsworth his life. Union and Confederate forces also fought in several other engagements as Lee's army retreated back to Virginia.

Various aspects of the Civil War continue to be debated endlessly, Stuart's expedition among the foremost. Stuart has been criticized for taking his troops out of communication with his commanding general; ironically, Kilpatrick has been criticized for *not* doing the same thing. Certainly some errors were made by many participants. But to an outside observer from a different era, an analysis is often problematic. Since we, from our modern day perspective, know the outcomes of various events, it is much easier to discern where the mistakes were made. But during the campaign and in the midst of battle, decisions had to be made with incomplete information, often under incredible stress.

The question has often been raised: Would the Confederates have won at Gettysburg had Stuart been present? This would seem to be the wrong question. Had Stuart been in communication with Lee throughout the campaign, likely a Battle of Gettysburg would not have occurred, at least as we know it. Before Gettysburg, the movements of Kilpatrick and Stuart had been, in a geographical sense, on the periphery of the campaign. But in a strategic sense, these movements had been absolutely central to the unfolding of the events. Without Stuart, Lee had been drawn into battle at Gettysburg largely because he lacked a clear knowledge of the movements of the Union troops.

Stuart's absence also caused other problems beyond a lack of intelligence of enemy locations. Particularly on July 2, various Confederate infantry forces had to fulfill roles that would normally have been assumed by cavalry. These deployments essentially took significant numbers of Southern foot soldiers out of attacks that would have benefited greatly by their presence.

To this day, arguments continue concerning various turning points of the Civil War. To pick a single event in such a vast conflagration is difficult, to say the least. History's tapestry is interwoven by a multitude of supporting and interconnecting threads; any war will be shaped by many social, political, and military events. Yet, the Gettysburg Campaign was a colossal event in the war. If it is not "the" turning point, it is certainly one of them. Lee entered Pennsylvania with approximately 75,000 fighting men. But his army would never be the same after the first week of July 1863; his forces were dealt a blow from which they never fully recovered. Many historians would argue that the Confederate states had a chance to win independence only as long as the Army of Northern Virginia was a viable fighting force. If one accepts that premise, then the consequences of Lee's defeat in Pennsylvania and the casualties taken by his army assume an immense importance.

The most cited reason for Gettysburg being the turning point is that from this point on, Lee was on the strategic defensive; his army would never again make a major invasion of the North. To use a present day sports analogy, it could be argued that before July of 1863, Lee was fighting to win; after that point, he was fighting "not to lose."

Meanwhile, in the Western Theater, the capture of Vicksburg, Mississippi was the culmination of another critical campaign. Not only did the Union now control the Mississippi River itself, but a vital Confederate supply line across the river to the Western states had been shut down. The combined physical, psychological, and strategic impact of the events in Pennsylvania and Mississippi was immense. Many important campaigns took place in the war, but after Gettysburg and Vicksburg, the Confederacy was consistently losing numbers and territory.

In June of 1863, a series of events, circumstances, and decisions had kept Stuart away from Lee. As one of those incidents, the engagement at Hanover, Pennsylvania, on June 30, had ramifications that went far beyond the scale of the fighting itself. To many historians, indulging in "what-if" scenarios is often considered outside the realm of professional boundaries. Yet in some cases, the consequences of decisions and events are profound enough to almost demand some after-the-fact speculation.

Stuart's decision to ride to Hanover turned out to be a fateful one. Had he waited a few hours while scouts monitored Kilpatrick's movements, some interesting possibilities might have been presented. Possibly the Southern Cavalry would have been able to pass through Littlestown to Gettysburg. Or, if Kilpatrick had turned back to seek a battle, Stuart might have been trapped between enemy cavalry and infantry.

When Confederates spotted the Union rear guard southwest of Hanover, the Southern officers could have let it pass. But their aggressive response triggered events that added several miles and hours to their movements. Stuart's decision not to destroy the captured wagons was also critical; it elongated his column to a considerable degree and forced him to keep a screening force at Hanover much longer than otherwise necessary.

What can be safely said without speculation is that the meeting at Hanover, and the result of the battle, forced Stuart farther away from possible communication with Confederate infantry. The encounter also added several more hours and miles to his movements. The events of June 30 greatly reduced the chances that Stuart would be able to reestablish communications with Lee and screen the marches of the Confederate infantry at a critical time of the campaign. Hanover helped ensure that the "information blackout" that Lee was under would continue for the next several hours, even as his forces were moving toward an unintended battle at Gettysburg.

Historians have occasionally debated whether the events at Hanover merit the term "battle." Some have argued that, considering the number of casualties, "skirmish" or "action" would be a more appropriate term. Yet casualties alone cannot be considered the only criteria for the importance of a military engagement. The conflict at Hanover involved both mounted and dismounted combat, artillery fire, and the presence of a large wagon train. The action took place over several square miles and involved decisions made by numerous high-ranking officers, including generals. Considering these logistical and tactical aspects, it is hard to see how Hanover could be considered anything less than a battle.

The final word, however, should be given to the commander of the Army of the Potomac. On July 1, 1863, Gen. George Meade sent a dispatch to Washington in which he stated, "Our cavalry, under Kilpatrick, had a handsome fight yesterday at Hanover."[318] That "handsome fight," along with many other events, helped shape a critical campaign of the war.

APPENDIX A
CONFEDERATE REQUISITIONS, CIVILIAN DAMAGE CLAIMS

On June 21, 1863, General Order No. 72 was issued from the headquarters of the Army of Northern Virginia. This document set forth regulations for the gathering of supplies by Lee's army while on Northern soil. The full wording is as follows:

> While in the enemy's country, the following regulations for procuring supplies will be strictly observed, and any violation of them promptly and rigorously punished.
>
> I. No private property shall be injured or destroyed by any person belonging to or connected with the army, or taken, excepting by the officers hereinafter designated.
>
> II. The chiefs of the commissary, quartermaster's, ordnance, and medical departments of the army will make requisitions upon the local authorities or inhabitants for the necessary supplies for their respective departments, designating the places and times of delivery. All persons complying with such requisitions shall be paid the market price for the articles furnished, if they so desire, and the officer making such payment shall take duplicate receipts for the same, specifying the name of the persons paid, and the quantity, kind, and price of the property, one of which receipts shall be at once forwarded to the chief of the department to which such officer is attached.
>
> III. Should the authorities or inhabitants neglect or refuse to comply with such requisitions, the supplies required will be taken from the nearest inhabitants so refusing, by the order and under the directions of the respective chiefs of the departments named.
>
> IV. When any command is detached from the main body, the chiefs of the several departments of such command will procure supplies for the same, and such other stores as they may be ordered to provide, in the manner and subject to the provisions herein prescribed, reporting their action to the heads of their respective departments, to whom they will forward duplicates of all vouchers given or received.
>
> V. All persons who shall decline to receive payment for property furnished on requisitions, and all from whom it shall be necessary to take stores or supplies, shall be furnished by the officer receiving or taking the same with a receipt specifying the kind and quantity of the property received or taken, as the case may be, the name of the person from whom it was received or taken, the command for the use of which it is intended, and the market price. A duplicate of said receipt shall be at once forwarded to the chief of the department to which the officer by whom it was executed is attached.
>
> VI. If any persons shall remove or conceal property necessary for use of the army, or attempt to do so, the officers hereinbefore mentioned will cause such property, and all other property belonging to such persons that may be required by the army, to be seized, and the officer seizing the same will forthwith report to the chief of his department the kind, quantity, and market price of the property so seized, and the name of the owner.
>
> By command of General R. E. Lee:
> R. H. Chilton, Assistant Adjutant-General[319]

Historian Edwin Coddington wrote a great analysis on the practical problems involved in enforcing General Order 72 and the effects of the Confederate presence in Pennsylvania. According to Coddington:

> Although these regulations were designed to prevent lawless confiscation of property, they naturally gave civilians no real choice in the matter of seizures. ....For compensation the officers were to offer payment in Confederate paper currency, which was greatly depreciated in the South and nearly worthless in the North, or as an alternative, a claim on the Confederate government to be honored in the future. For the farmer or merchant such unprofitable transactions were to be avoided by all means; hence their anxiety to hide their valuables. The only difference between illegal and legal confiscation was that the latter gave civilians the possibility of recovering some of their losses should the Confederacy win the war.
>
> Under the circumstances the citizens of Pennsylvania could not have had a fairer arrangement to compensate them, but the primary purpose of these regulations was not a humanitarian one. Lee wanted to make sure that seizure of the region's movable wealth was done efficiently and for the benefit of the whole army. He also knew that wanton and indiscriminate pillaging and destruction of property by individual soldiers would break down discipline and reduce the effectiveness of the army....
>
> Basically the regulations demanded the impossible, since they resulted from certain questionable assumptions in their formulations. A correspondent of the *Richmond Sentinel* saw some of their inherent difficulties when he said that "the doctrine of not using or destroying some of the private property of an enemy while in his country is pure abstraction. You cannot possibly introduce an army for one hour into an enemy's country without damaging private property, and in a way often in which compensation cannot be made...."
>
> Another difficulty was that a literal interpretation of the regulations prohibited the supply chiefs from delegating authority in the matter of seizures. They themselves could not cover an area large enough to secure all they needed, yet when underlings took over the job there were many more opportunities for abuse of power and unwarranted confiscation of property. Such instances occurred.
>
> Confederate officers policed the activities of their men effectively in town, but not so well in the country. The infantry, easier to restrain than the cavalry, were "kept close in ranks and marched slowly," and when camped near a town they could not get in without a pass. On the march the cavalry kept in front and on the flanks and thus had better opportunities for plunder of property and seizure of horses. Some of the troopers traveled in small detachments, scouring the byways and back roads and visiting isolated farms....
>
> ....Southerners mentioned not only thefts of horses, but other kinds of property as well. In looking for food and fuel, privates often ignored Lee's orders and took fence rails, whiskey, chickens, and other kinds of food in large quantities. ....Considering the great scarcity of the necessities of life in the Confederacy, such depredations by hungry and footsore soldiers were understandable....
>
> Although many a Rebel plundered and stole with great efficiency, the army as a whole never got out of hand. Although most Southern soldiers would have liked to indulge themselves at the expense of their rich and hated enemy, nevertheless they exercised restraint out of respect for their commander and the code of the times. ....Because of the North's greater wealth and more efficient transportation facilities the Army of the Potomac had not made it a practice to live off occupied territory, so that a systemic and thorough stripping of an area's resources to support an army was a strange and horrifying experience to the Union soldier. His unsophisticated mind could see little if any difference between legal seizures of goods by duly constituted military authorities, who paid for their confiscations with worthless

money, and brazen theft with no pretense of legality. For that matter, neither could the shopkeepers and farmers in Pennsylvania.

Favorable comment on the humaneness on General Lee should not obscure the fact that the Confederates were waging war against the United States, and in June of 1863 they extended it to Pennsylvania. For many Pennsylvanians the war now attained the same dimensions of senseless cruelty as it had come to have for their former compatriots in Virginia.[320]

Certainly the economic disruption to the south central counties of Pennsylvania was immense. In the 1860s and 1870s, the General Assembly of Pennsylvania passed several acts by which citizens of these counties could apply for compensation for damages that occurred from troop movements, seizures, etc.

These Acts were:
- Act of April 16, 1862. (Initial Act.)
- Act of April 22, 1863. (Stuart's Raid of 1862.)
- Act of Feb. 15, 1866. (Burning of Chambersburg.)
- Act of April 9, 1868. (First general commission to examine all claims.)
- Act of May 27, 1871. (Additional relief for Chambersburg sufferers.)
- Act of May 27, 1871. (Second commission to revise all claims, grant awards, and issue certificates.)

One can get a sizable headache trying to follow the trail of bureaucratic red tape and political wrangling that resulted from these legislative acts, at both the state and federal level. The state of Pennsylvania initially approved the idea for damage payments to citizens. But when it became apparent how much in damages was to be paid, government officials were appalled, so Pennsylvania sent commissioners to verify all the evidence. By 1872, the state had issued certificates to individual claimants for specific amounts. In theory, those amounts were to be paid once the act was passed to approve the funds. Later, the claims were pushed up to the federal level, (Quartermaster Dept.) but as the federal government would only consider payment for damages done and items taken by Union soldiers, any claim concerning Confederate damages was sent back to Pennsylvania. The state act to approve the funds was never passed and the money never paid. (The United States Government did pass legislation to approve funds for the federal claims, although the great majority of claims were rejected.)

The evidence filed in the claims process can provide interesting information on individual citizens and some of their encounters with soldiers. Also of great value is that several maps of that era show the locations of houses along with the name of their owners. When the accounts in the claims are referenced to the maps, the information can be used to actually trace the movements of troops, particularly along secondary roads by units whose marches have, in some cases, been "lost" to history. (See the bibliography for the maps used in this study.)

APPENDIX B
COMPANIES OR TROOPS?

The Union cavalry regiments that fought at Hanover were, on average, larger than those in other Union cavalry divisions. Even with various companies being on detached duty, the typical Union regiment in Kilpatrick's command numbered 500-600 men. Of the Federal forces that fought that day, the following generalizations can be made:

   1 Union cavalry regiment = 3 battalions
   1 Union cavalry battalion = 2 squadrons
   1 Union cavalry squadron = 2 companies (or troops)
   1 Union cavalry company = approximately 50-60 men.

The Union units at the battle were also larger, on average, than their Confederate counterparts. Most Southern companies at Hanover numbered approximately thirty to forty men. Since the typical Confederate regiment contained ten companies rather than twelve, their regiments numbered generally between 300-400 soldiers.

Although the cavalry companies were often informally called "troops" in the Civil War, "company" was still the official designation at that time. These units were not formally called "troops" until 1883.[321]

APPENDIX C
PENNVILLE, BUTTSTOWN, MUDTOWN, MUDVILLE, OR DRECKSTHEDDEL?

In 1863, a few houses were located near the intersection of the Westminster and Hanover-Littlestown Roads. Well after the war, this village acquired the name Pennville. In the 1876 Atlas of York County, the village is named Buttstown, but this term appears to have been rarely used and for only a short time after the war. In the John Gibson *History of York County, Pa* (1886) and the George Prowell *History of York County, Pennsylvania* (1907), neither historian mentions the village ever being called Buttstown. Even in a short biography of John Butt, (whose name presumably is how the title Buttstown originated), Gibson's publication states that Butt lived in Pennville.[322] Prowell also wrote a 1919 article in the *Hanover Record Herald* on the history of Pennville; in that article, the name Buttstown is never mentioned.[323] A later historian, William Anthony, said that the village was formerly known as "Buttstown, Mudtown, or Dreckstheddel."[324] Robert G. Carter marched through the village on July 1, 1863, as an infantryman with the 22nd Massachusetts. Carter said, "The small clump of houses near where the Westminster Road comes in from Union Mills to the south (the road we marched in on) was called 'Mudville' during the war; it now enjoys the much more dignified name of 'Pennville.' "[325] As no title of any type is listed on the 1860 York County map, it seems unlikely the village had an official name in 1863, so perhaps Mudville or Mudtown was the most commonly used designation at that time.

APPENDIX D
WHICH 1ST WAS FIRST?

Various books and articles have differed as to whether the 1st West Virginia or 1st Vermont led Farnsworth's brigade that morning. When interpreting the Union line of march, most recent writers have cited either *Encounter at Hanover* (1962) or *The Battle of Hanover* (1945), both of which place the 1st Vermont at the front of the brigade. Neither of these works is footnoted, but each copied its sequence of regiments from the writings of George Prowell in his *History of York County,*

Pennsylvania (1907). Prowell's order of march appears to have been based on accounts of Lt. Henry Potter of the 18th Pennsylvania; several of Prowell's sentences are almost word-for-word reprints of that officer's accounts. However, since Potter commanded the rear guard, he could not have possibly witnessed which regiment entered Hanover first. Whether he saw the order of march before that point also seems unlikely. He listed the 5th and 6th Michigan near the front of the column when those two units were not even with the main body of the division at that time. Essentially, Potter's sequence of regiments was adopted by local writers who never questioned its merits, and the information was later repeated in other secondary works.

It would seem that an examination of the accounts of the Vermonters and West Virginians themselves would settle the issue, yet contradictions can be found even among the men in those two regiments. In 1890, Capt. Henry Parsons (1st Vt.) wrote, "...the 1st Vt. was in advance, the 1st [West] Va. and 5th N. Y. were in the center, and the 18th Pa. was in the rear."[326] Yet in direct opposition to this statement is the evidence found in the official report of Maj. Charles Capehart of the 1st West Virginia, dated August 17, 1863, which says that "The First West Virginia was in the advance of the brigade, and had passed through the town," when the attack started.[327] (On many occasions in Civil War reports, letters, diaries, etc., the word "advance" was used as a noun rather than a verb, an indication of being in the front of a movement.)

But the writings of Parsons, Charles Blinn, Horace Ide, and S. A. Clark, all of the 1st Vermont, have one important aspect in common; they each refer to actually being in the town when the first shots were fired. According to Clark, "the head of the column was nearly through the town" when firing was first heard.[328] Ide stated that when the first shots were fired, the column continued forward for a time, and "just as we were near the toll gate we saw the enemy's column appear with a battle flag at the head."[329] This tollgate was located on the Abbottstown Road (Broadway), very close to the Ridge Avenue intersection. This statement indicates the vanguard of the Confederate charge reached well past the square. If the 1st West Virginia had been behind the 1st Vermont in the line of march, it would have been almost impossible for the Confederate assault to reach this point. (The 5th New York had withdrawn off the street by that time; but the accounts of the 1st West Virginia indicate they passed through Hanover before they deployed.)

Another aspect to consider is that two companies (D and M) of the 1st Vermont were involved in direct counterattacks against the onrushing Confederates; even various accounts of 5th New York men note their participation. It seems much more likely that the troops closest to the action would have the most involvement. (The 1st West Virginia, when it raced back through the town, seems to have been involved more in a "mop-up" capacity at that point. The casualties they sustained could have occurred almost anytime that day considering they were involved in skirmishing for several hours.) An 1887 article in the *St. Albans* (Vermont) *Daily Messenger* states that the 1st Vt. was dismounted in the town when the first shots were fired and that "the Fifth New York, the first in rear of the First Vermont Cavalry, received the full force of the shock, which extended some ways into the First Vermont Regiment."[330]

Apparently, the two regiments were not far apart, no matter in what order they were marching. To further muddy the waters, one 1st West Virginia man, J. P. Allum, also referred to being in the town when the first shots were fired.[331] It is possible that any of these men were confusing the small village of New Baltimore as being part of Hanover. But the overall weight of the evidence seems to indicate that the 1st West Virginia most likely led Farnsworth's brigade that morning.

APPENDIX E
CONFEDERATE ARTILLERY LOCATIONS

Several primary accounts refer to Southern artillery in the early stages of the fighting but do not specify a precise location. A good example is Henry Sell's statement that, "A Confederate cannon planted on the Keller farm to the south fired the first shot of the battle".[332] The Sell and Keller properties shared a common border, but Keller owned 161 acres of land in 1863, and Sell did not state the specific spot.

Another aspect that makes an interpretation difficult is that the situation at the start of the battle was a very fluid one. The positions of the guns were changed at least a few times, and their exact locations at any given moment are difficult to pinpoint. Henry Matthews, an artilleryman with Breathed's battery, wrote a postwar account of his unit's involvement in the battle. Matthews' writings indicate that when the fighting started, his battery was very close to the action. It is possible that when the first shots were fired, Confederate artillery was situated on the ridge where the Rest Haven Cemetery is now located. (But if so, possibly on the opposite side of the Westminster Road from the cemetery.) Matthews stated that when the Union troops counterattacked, the Confederate artillerymen had to *fall back* to the more significant hill(s) which then became a portion of their dominant line of battle. It was from this more prominent high ground that the great majority of Confederate shells were fired.

In his 1907 *History of York County, Pennsylvania*, George Prowell wrote that the Confederates "planted two cannon on the Samuel Keller farm, near Plum Creek, and two on the Jesse Rice farm, along the Westminster Road."[333] This sentence has had a huge impact on later writings; it has been the basis for almost every interpretation that has followed concerning the Southern artillery positions. His statement (and much of his other work) was later included word-for-word in the books *Battle of Hanover* and *Encounter at Hanover*. I believe that Prowell's general concept of the locations of these guns is likely accurate, *in that Confederate guns were certainly on these two tracts of land*. But later interpretations of this statement have introduced some misconceptions.

Since Prowell was relating events for a local audience in the early 1900s, he often used references that those people would recognize, thus his use of the name Jesse Rice. But according to tax records, deed documents, etc., Jesse Rice did not live in this area until at least 1865, and did not own any land until 1872. (See below for citations concerning the ownership of this tract of land.) Later writers assumed that Rice was the owner during the 1860s and his name became associated with an area in which he had no actual Civil War connection.

Most, if not all, of the secondary accounts that mention Rice's name were written in the early 1900s, and the actual term "Rice's Hill" seems to have originated much later. Local historian Robert E. Spangler drew a map dated June 30, 1936, which used this term.[334] He also used that name in an account included in the 1945 book *Battle of Hanover*.[335] With the publishing of *Encounter at Hanover* and other works, the association of Rice with the area became unquestioned and "history" was made.

Another difficulty was introduced in 1963, with the publication of *Encounter at Hanover*. Included in the book was a map that had Confederate artillery drawn near a bend in the Plum Creek, close to the Keller house, likely based on the earlier statement of George Prowell. The details of this map have been recopied so many times in books and articles that they have been accepted as fact. However, no primary source evidence has been found which indicates that Confederate guns ever fired from this lower ground along the creek. From there, their field of fire would have been close to non-existent toward almost any Union position. The ridge to the north of this low ground, where the

Rest Haven Cemetery is now located, would have prevented any line of sight toward Hanover. A view toward much of the Littlestown Road, a critical route on the Confederate left flank, was also blocked from this position. But locations a few hundred yards away, in almost any direction, would have given them a much better field of fire.

Of the available primary sources, several have indications in common; the Confederate artillery was on high ground in dominant positions and, at least initially, was on or very near the Westminster Road. (Up to this point I have located ten actual primary source accounts that mention the Confederate artillery fire in the early stages of the battle. Three do not mention any specific terrain features; the other seven all state that the Confederate guns were positioned on "hills", "high ground", etc.)

It is almost certain that artillery was positioned on the land that Prowell referred to as the Jesse Rice farm. But since it was actually owned in 1863 by Jacob Forney, perhaps a better name would be "Forney's Hill." This tract actually contained two areas that offered prominent elevations. One is the hill at the intersection, and immediately south, of the Westminster Road and Cooper Road. Any artillery on this hill east of the Westminster Road would likely have been on Forney's property, while any guns west of the road were on land owned by Samuel Keller.

The other area is the high ground several hundred yards to the northeast of the last mentioned site. (A portion of this hill is covered by the easternmost section of the South Hills Golf Course.) Jacob Forney's boundaries, however, did not completely cover this entire elevation; artillery could have been positioned on the eastern portion of the hill on land owned by Jacob Forry. This terrain dominates the town and was almost certainly a Confederate artillery position, likely after they fell back as indicated in Henry Matthews' account.

The following documents provide information regarding the ownership of the Forney/Rice parcel:

1-York County Deed Book 6J, page 582. According to this deed, Jacob Forney died in 1872 without a will. His executors, Jesse Forney and Ephraim Ritter, applied to the Court on August 31, 1872 to put the land up for public sale. Jesse Rice was the highest bidder, and the result of the sale was confirmed by the Court on September 6, 1872. This same parcel of land passed through other later owners including Louis Dellone and D. E. Winebrenner. A portion of this tract is presently the section of the South Hills Golf Course situated east of the Westminster Road. This parcel has a very distinctive shape, and when one maps out the land according to the measurements given in the above deed, the borders are still apparent, particularly when examining aerial photographs.

2-According to Paradise Township tax assessment books, Jesse Rice is listed as living in that township from 1860-1864, inclusively. It can be confirmed that this man is the same Jesse Rice who later lived south of Hanover by comparing the census data from Paradise Township, 1860, with Heidelberg Township, 1870. (Our modern Penn Township was part of Heidelberg Township at that time.) In these two census reports, the names and ages of Rice, his wife Caroline, and his daughters clearly indicate the same family.

3-Heidelberg Township tax assessment books. The tax books support the deed cited above. According to these records, no Jesse Rice resided in Heidelberg township until at least 1865. More importantly, the tax books confirm that Rice owned no land until 1872.

4-In the 1870 federal census, Jesse Rice is listed as having $4,000 of personal estate, but no real estate. Interestingly, two years later, Rice purchased the land for just more than $4,824.

5-In the septennial census of Heidelberg Township taken on November 19, 1863, no Jesse Rice is listed as being a tax paying citizen of Heidelberg Township. Jacob Forney (farmer) and Jacob Forney (retired tanner) are both listed.

APPENDIX F
ATTACK ROUTE OF THE 2ND NORTH CAROLINA

The "outside street", described by Payne in his letter to Fitzhugh Lee, was likely what is now known as Exchange Place, the alley parallel with, and south of, Frederick Street (between Frederick Street and Walnut Street).[336] This lane is shown on both pre-war and postwar maps of the borough. Payne's letter clears up some difficulties in interpreting this part of the battle. The 2nd North Carolina movement has often been described as a flank attack through the crop fields southwest of town, but this area contained some terrain features (Plum Creek, fences, and hedge rows) that would have restricted their approach. The time and effort necessary to bypass these obstacles might have given Union forces on Frederick Street time to react to that threat. But by reaching this "outside street" before making an all-out charge, the 2nd North Carolina would have bypassed these obstacles and then had an unobstructed path to make their flank attack. Most likely Payne's troops used Westminster Road to get past Plum Creek. It is possible they then rode a portion of Frederick Street before moving onto the "outside street."

In various secondary publications, the attack of the 2nd has sometimes been referred to as a two-pronged attack, with part of the regiment actually charging along Frederick Street. Once again, writers have based this opinion on the books *Encounter at Hanover* or *The Battle of Hanover*. These two works took their core battle narrative, in most cases word-for-word, from Prowell's *History of York County, Pennsylvania*, which first introduced this interpretation. Strangely enough, although Prowell conducted a few interviews with North Carolina officers involved in the battle, none mentioned a two directional attack by *that regiment alone.* In fact, Lieutenant Colonel Payne, who led the charge, stated more than once that he personally directed his troops onto the Union flank *before they made contact with the enemy.* Yet Prowell later wrote that Payne led a portion of the regiment in a charge along Frederick Street. Some possibilities exist that may explain the discrepancy. The two-pronged attack may refer to the fact that the Virginians were pressuring the Union rear at the same time that the 2nd North Carolina charged upon their flank. Another possibility is that various companies of the 2nd used different side streets to strike the Union column. This would appear to be a multi-columned strike, even while the entire regiment made its approach from the same direction.

The strength of various Confederate units in the Gettysburg Campaign has long been debated. Company muster rolls, when available, are generally considered to be the most accurate representation of the numbers of men available. John W. Busey and David G. Martin undertook an extensive study of these sources. Their work was published in their book *Regimental Strengths and Losses at Gettysburg*, considered an indispensable resource for students of the campaign. As their research has shown, the problems of judging exact numbers in the Southern units are substantial, largely because so many of the June 30, 1863, company muster rolls are missing for the Army of Northern Virginia. As Mr. Busey and Martin state, "The primary cause for the lack of complete rolls for the army is the destruction or loss of large numbers of records both during and subsequent to the war. Secondly, many Southern units, owing to the stresses of the campaign, simply did not take the time to complete a formal muster when called for on 30 June. This is evident from the large number of rolls made out for the period of 30 June which are dated in late July or August 1863. The overwhelming majority of these rolls contain post 30 June data and represent the status of the unit at the time the report was finally filled out, not that as of 30 June".[337] The authors go on to say that of the 277 companies that comprised the Southern cavalry in the campaign, they were only able to locate 58 of the June 30 muster rolls in the National Archives.

Clearly, the problems in citing exact figures for these units are immense. The strength and casualty figures of the 2nd North Carolina, in particular, are a real enigma. The following accounts reveal the wide range of numbers that various sources have listed for the regiment on June 30, 1863:

A. A report written by Capt. William A. Graham, dated October 1, 1863 states, "In this affair [Hanover] we lost twenty-two men, two officers, out of fifty carried into action."[338]
B. In an enclosure from Stuart, separate from his primary campaign report, the 2nd North Carolina is listed as having a total of sixty men killed, wounded, or captured for the entire expedition. (Note: Evidence suggests that the 2nd North Carolina sustained only one casualty at Gettysburg itself.)[339]
C. Shortly after the war, William H. Payne wrote a collection of observations entitled "Notes on War and Men - Summer 1865." In reference to the fight at Hanover, he stated "...156 entered the town - 23 only got out - the rest were nearly all killed or wounded."[340]
D. In a letter from William H. Payne to Fitz Lee dated May 12, 1871, he wrote, "...our hurried march had prostrated men and horses; and when we got in front of Hanover I certainly had not more than 100 effective men in my regiment..."[341]
E. The following was included in a local newspaper article. "Captain W. A. Graham, who commanded Company A of the Second North Carolina Cavalry, has communicated to the Historical Society of York County some interesting facts, heretofore unpublished, relating to the cavalry engagement at Hanover in 1863. He says that his regiment, by order of General Stuart, the Confederate commander, led the second charge into the town after the Virginia regiment had been repelled with considerable loss. This statement corresponds to that made two years ago by Col. Paine who commanded the regiment and was captured near Winebrenner's Tannery at the end of Frederick Street. Col. Paine and Capt. Graham, both of whom are still living, estimate the loss to the Second North Carolina in this charge in killed, wounded, and captured at two hundred and fifty officers and men."[342]
F. George Prowell wrote that "This regiment was commanded by Col. Payne.... His regiment lost the heaviest in this fight; it was claimed by himself in a personal statement to the writer that his loss must have been nearly 200; many of these were captured..." These last two accounts were apparently the basis for Prowell's interpretation in his *History of York County* in which he believed the strength of the 2nd North Carolina to have been "nearly 500 men".[343]
G. In the study *Regimental Strengths and Losses at Gettysburg,* the authors cite several muster rolls of various companies in the 2nd North Carolina. Two of these documents are dated June 30, 1863, while two others are dated July 1. The rolls of just those four companies alone (Companies B, C, G, and H) total 191 troops.[344]

Interestingly, Graham's statements, which vary the greatest in the dates they were given, also vary the most in the numbers given. But both his high-end and low-end estimates would appear to be well off. In Kilpatrick's report on the battle, he admits that the Confederate losses were somewhat less than his own. In light of that statement, it is difficult to see how the highest figures cited in the Hanover newspapers could be correct. Also, the statements given to Prowell occurred almost four decades after the earlier accounts and were likely not as accurate. If Prowell's article was correct, the 2nd North Carolina would have been considerably larger than any other regiment in the brigade. Considering the losses the 2nd took in earlier battles in the campaign, this would appear to be extremely unlikely. But the lowest numbers cited by Graham also seem to be well off the mark; the Confederate official report enclosure lists more men as casualties than Graham's report claims were actually in the battle.

However, the discrepancies between the muster roll figures and Lieutenant Colonel Payne's pre-1900 statements are puzzling. One likely explanation is that some men listed on the regimental rolls were actually on foot because of horses that had dropped from the ranks; a significant number of the 2nd North Carolina may have been in this situation. If so, they could have been present at Hanover and taken part in dismounted skirmishing without being in the mounted charge. Or, some may have fallen behind on the march and never caught up to the regiment. The letter from Payne to Fitz Lee mentions the losses from the severe hardships of the movement. Other accounts confirm that situation. In the *History of the Ninth Virginia Cavalry*, Colonel Beale wrote that squads were sent on the morning of June 30 to "collect horses for our dismounted men..." Considering the casualty figures given by Stuart and the earlier statements from Payne, it is my belief that the 2nd North Carolina had close to 150 men in its mounted attack.

APPENDIX G
ELIZABETH SWEITZER WALTZ

On June 30, 1896, the *Evening Herald* newspaper ran an article entitled, "The Battle of Hanover," in which it stated that "...not a single citizen was hurt..." during the fighting. A few years later, however, an interesting development took place that caused the same newspaper to reconsider this claim and to publish another article on March 1, 1899, titled, "A Pension for Mrs. Waltz." According to this later piece, Mrs. Lizzie Waltz had visited Hanover in February 1898 "looking for evidence to substantiate her claim for a pension." As a young girl in 1863, Lizzie had been employed as a domestic servant to the family of Rev. M. J. Alleman. The *Evening Herald* states that "Mrs. Waltz, then Lizzie Sweitzer, alleges that she was one of the girls on the street, feeding the soldiers, and that before she was able to reach shelter, she was struck on the ankle by a carbine or pistol ball, inflicting a serious wound, from which she never fully recovered. At the time of her last visit here she walked with a crutch as the result of the wound." The article goes on to say that "From the fact that an act for her relief has been passed, she must have been able to substantiate her allegations to the satisfaction of the Committee on Pensions in Congress."

From the wording in this article, one could infer that the *Evening Herald* was skeptical of this claim, and with good reason. It was then approximately thirty-five years since the battle, and no account had surfaced up to that point of any civilian injured during the fighting. Yet Elizabeth Sweitzer Waltz had not only been granted an invalid pension, but a bill had been introduced in the United States House of Representatives to facilitate the matter. Although the full story may never be known, some interesting documents can be found in the pension file itself.

According to information found in the National Archives, the birth date of Elizabeth Sweitzer was July 16, 1848, which indicates that she was 14 years old during the battle.[345] (In the 1860 federal census, her age is listed as seventeen. If that information is correct, she would have been nineteen or twenty at the time of the battle.) Eventually Lizzie met Levi Waltz, and the two were married on Sept. 15, 1868 at Reedsburg, Wisconsin. Sometime after that, the couple and a son moved to Watertown, South Dakota. Levi had served in Company D of the 19th U. S. Infantry and was granted a federal invalid pension beginning in June 1886. After he died on March 18, 1890, Elizabeth applied for a pension as his widow.[346]

Elizabeth had been receiving payments of $12 per month on her widow's pension, then decided to apply for an invalid pension. Apparently, the application caught the attention of the right people. It is possible that Lizzie appealed personally to a congressman to work on her behalf. On February

10, 1898, a bill was introduced in the United States House, and was later referred to the Committee on Invalid Pensions. Within a few months, the Committee recommended that the bill be passed and included a report to accompany the legislation.

Excerpts of the report include the following:

> Evidence filed with this committee shows that on or about July 1, 1863, at Hanover, Pa., claimant was on Frederick Street, circulating around the soldiers, giving food.... While so doing a detachment of rebel cavalry charged on them, and in the charge she was either struck with a bullet in the left ankle or it was cut by a shoe of a horse. She was knocked down, trampled upon by the horses, and lost consciousness. She was pulled out from under the horses by a Union soldier and taken to a house where she was attended by a physician. She states that her wound required the frequent attendance of a physician and a nurse for fifteen years following its incurrence and she was obliged to use crutches or a cane to walk, not being able to bear her weight upon the limb. Nor is she now able to. She has been obliged to keep it bandaged continually and to wear splints most of the time to keep it in place, and it requires more than her pension to pay doctor's bills and for medicines.
>
> She is unable to perform any manual labor for her support. Her husband was an invalid for two years before his death, requiring constant medical attention, so that she was compelled to sell her home and farm to pay for the same, and at the present time has no property nor income except her pension.
>
> A photograph of the injured member has been filed with the committee for inspection. The women who served as nurses to our sick and wounded soldiers and those who went to the battlefield and fed them when contending against the enemy and braved the dangers and received wounds while so engaged are entitled to the highest consideration of this Government. The bill is reported back with the recommendation that it pass when amended as follows: Amend the title so as to read, "A bill granting an increase of pension to Lizzie Waltz."[347]

Unfortunately, the pension records in the National Archives leave as many questions as answers. The file does contain the documents used to uphold her application for the *widows* pension. However, what is missing is the actual "evidence filed" to support her *invalid* pension. Which Hanoverians, if any, gave testimony in this case is not stated. Furthermore, no other accounts have surfaced that would indicate whether any locals were wounded. It is peculiar that if a person was incapacitated immediately after the battle, it would not have caused comment and become known by local newspapers. Possibly the injuries caused by the battle were minor and then became more troubling over several years, crippling Lizzie as she got older. At the very least, it appears that Lizzie was helping feed the Union soldiers that morning and for that should be treated as a heroine of the battle.

APPENDIX H
JOHN HOFFACKER

One of the most enduring stories of the battle concerns John Hoffacker of Company E, 18th Pennsylvania. According to almost every book or article on the battle, this soldier was born and raised in West Manheim Township and left his local farm to enlist just two months before he was killed. But most, and possibly all, of this statement is incorrect. Fortunately, in this case the real story is much better than any myth, and it is past time to give full tribute to this soldier and his family. The history of the Hoffacker family is an amazing tale of heroism, perseverance, hardship, loyalty, and tragedy.

Since the typical interpretation of John Hoffacker has been so unquestioningly accepted, it is worth taking a look at how and when it came into the public consciousness. On July 9, 1863, the *Hanover Citizen* included a letter from Dr. Perrin Gardner, the surgeon in charge of the care of the wounded. In that report, John Hoffacker's name was listed among the men killed in the battle.[348] About ten years later, another *Hanover Citizen* article mentioned Hoffacker as a casualty.[349] But by 1886, an interesting anecdote was added. In the Gibson *History of York County, Pa.*, George Prowell wrote a section on the battle in which he referred to the original report of Dr. Gardner. But inserted in the surgeon's casualty list was a notation, presumably written by Gibson or Prowell, that John Hoffacker "lived a few miles south of Hanover."[350] (This statement is not in the surgeon's original report and appears to be the first reference to that aspect of the story. *In neither previous article was Hoffacker noted as having any local ties*. It is almost inconceivable that a soldier could be killed that close to his home without major press coverage of the event in the months immediately following the battle.) This vague reference to the location of the family home is apparently the beginning of the typical John Hoffacker interpretation. That concept was repeated about a decade later in another very ambiguous citation. According to *The Evening Herald* of June 30, 1896, Hoffacker "was from below Hanover."[351] It could be argued that *The Evening Herald* was technically correct. But by 1907, an added dimension was introduced with the publishing of Prowell's *History of York County, Pennsylvania*. According to Prowell, Hoffacker lived in West Manheim Township at the time of his enlistment. It should be noted that up to that time, any references made to the location of the Hoffacker family home were vague at best. But Prowell's 1907 statement was copied in *The Battle of Hanover* (1945) and *Encounter at Hanover* (1963), and then repeated by multiple writers who never questioned its accuracy. Because of endless repetition of information based on the same secondary source, "history" was being made.

But the true story can be found in the actual words of the Hoffacker family members. Documents found in one Federal pension file, along with various compiled service records, reveal much of the family's history in the mid 1800s. These documents include sworn affidavits from the soldier's parents, siblings, friends, and neighbors, along with letters written by both John and his brother, William. Census and tax records also confirm every statement made by the Hoffacker family.

The family did have Pennsylvania ties. Henry, John's father, was born in Pennsylvania, and after he married Elizabeth in 1833, the couple may have resided in York County for a time. (According to William's military service record, he was born in York, Pennsylvania. But this citation is in direct contradiction to the census records, which state that all the Hoffacker children were born in Maryland.) However, no Henry Hoffacker is listed in any Manheim Township Tax Assessment Book for the years 1837-1849. (At that time West Manheim Township was part of Manheim.)

From about 1850 on, the family's whereabouts can be firmly established. By that year, they were living in Maryland, which is confirmed by several pieces of testimony given by family members and neighbors and by the following information found in the 1850 federal census, Carroll County, Maryland, District Six, P. O. Manchester, Household #20, Family #20.

Henry M. Hoffacker   39 [years old]
Elizabeth   40
George E.   16
John   14

> Ann M. 12
> William 10
> Martin S. 8
> Martha A. 6
> Mary E. 6
> Frederick Cummings 21 - Laborer

In the late 1840s, an incident occurred that set in motion a critical chain of events. According to testimony given by Henry in a federal pension file, he sustained a rupture (likely in 1847). In the next few years, internal damage from the injury made demanding farm labor almost impossible. To support the family, Henry needed to find a less strenuous occupation. He sold his Carroll County home and bought the Gunpowder Falls Paper Mill in Baltimore County, Maryland. (The Hoffackers moved to Baltimore County in 1850, shortly after the above census was taken.) Henry operated the mill, while his children worked a piece of nearby farmland. Several individuals testified that John and William were active in working at the mill and helping with the Maryland farm. To help support the family, the two did not accept any payment but received room and board. According to sworn statements by the father, mother, sons, daughters and neighbors, the Hoffackers lived in Baltimore County throughout the 1850s. Their whereabouts in 1860 are also confirmed by the following information found in the 1860 federal census, Baltimore County, Maryland, District 6, P. O. Union Meeting House, Dwelling #182, Family #181.

> Henry Hofacker [sic] 50
> Elizabeth 51
> George 26
> John 21
> William 18
> Martin 16
> Martha 13
> Elizabeth 13
> Mary William 24 Domestic
> Michael 2

(A few men named John Hoffacker can be found in the 1860 census rolls of West Manheim and Manheim Townships. But the ages given for these individuals, along with the names of other household members, clearly indicate that they cannot be the soldier in question.)

The family's real problems began about 1859. By that time, Henry had lent a large amount of money to another man, who then defaulted on that debt. Apparently, the Hoffackers had no way to collect on the sum, and they fell into deep financial trouble, eventually owing money to a few different creditors. Their mill was sold at a sheriff's sale, but the proceeds did not cover all that they owed. Family members continued to work the mill under the new ownership.

In 1862, the mill was sold to yet another owner, and a different phase of the family's fiscal crisis began. Up to this point, the family had remained together through the hardships. But the most recent sale of the mill appears to have been a catalyst of events that would tear the family apart. In March of that year, William enlisted in the 3rd Maryland Infantry, possibly to earn a paycheck to help support the family. Meanwhile, John and George began working at a mill on Deer Creek in York County, Pennsylvania. (Deer Creek runs through Shrewsbury and Hopewell Townships.) Another son continued to work the family's Maryland farm, while the parents helped as much as they were able.

In September of that year, John enlisted in the 18th Pennsylvania Cavalry. From the time William and John joined the service, they consistently wrote home to their parents. (Some of these letters can be found in the federal pension file cited below; a few specifically mention the address of the family's Maryland home.) The faithfulness of the sons is very evident in these letters; they often included significant amounts of their pay to help their parents support the family. In one correspondence to his father, dated Dec. 28, 1862, John underlined the following passage for emphasis: "...if I fall you and mother is to have it all..." (in reference to his pay).

Sometime in 1863, the parents (and possibly a few other children) moved to Parkton, Maryland, where they began to operate a hotel. Although several siblings had become separated, it appears that the family fortunes may have been on the road to recovery, with the various members doing their part for everyone's survival. But on June 30, tragedy struck when John was killed at Hanover. Ironically, *if* John fell on Frederick Street, William was literally on the same road as he marched toward Littlestown with the Union Twelfth Infantry Corps. (Note: From 1850 to the time of John's death, his whereabouts can be firmly established; never did he or his parents own any land in York County, Pennsylvania, during this period. Also, John had been with his regiment approximately nine months by this time, not the two months often stated in various secondary works.)

After the Gettysburg campaign, the Union Twelfth Corps was transferred to the Western Theater. On or about November 18, 1863, near Stevenson, Alabama, William was charged with stealing $59.85 from another soldier in his regiment. (His service record does not state the outcome of this incident.) By this time, William would have learned of his brother's death and realized that the financial burdens on the family had now multiplied. Although it is impossible to know William's motivations, it is not hard to see how his emotional stress at this point could have led to an attempted theft.

Eventually, the 3rd Maryland was transferred back to the Eastern Theater. On May 12, 1864, William was wounded in the right knee at the Battle of Spotsylvania, Virginia. Documents on file from the surgeon general's office seem to indicate that his recovery prospects were good, but apparently he never fully healed. While on furlough, William died at his home in Parkton, Maryland, on February 3, 1865.

The parents persevered, even while their own health worsened considerably. In 1867, they took the savings they had made from keeping the hotel at Parkton, Maryland, and purchased property near Railroad, York County, Pennsylvania. Elizabeth, and later Henry, applied for a federal pension as mother and father of a deceased soldier(s) who had contributed to their livelihood. By March of 1891, Elizabeth had died, and her estate was consumed in paying debts. Henry, totally incapacitated by this time, lived with another son until he passed away.

The present day burial site of John Hoffacker in Mount Olivet Cemetery is well known. It is likely that the parents' financial problems were the reason the body was not removed from the Hanover area. After William died, Henry and Elizabeth decided they wanted the two sons buried side by side. Unfortunately, they did not have enough of their own money to properly bury William. In a terrible irony, they were forced to use the soldiers' back pay, along with William's reenlistment bounty, to pay for this final tribute. Later, Elizabeth and Henry were also buried at Hanover, where their remains rest beside their faithful sons, William and John.

(The above information can be found in the National Archives, Washington, D. C. Much of the evidence is from the Federal pension file under William's name, which was applied for by the parents (certificate number 287,744). This file includes literally dozens of testimonials and statements to support the parents' claim. Other supporting information can be found in the compiled service records of John Hoffacker (Company E, 18th Pennsylvania Cavalry), and William Hoffacker (Company A, 3rd Maryland Infantry).

APPENDIX I
LOCATION OF THE SCHWARTZ SCHOOLHOUSE FIGHT

Colonel Gray's report on the actions of the 6th Michigan included the following; "On approaching the last named place [Hanover] we came upon the enemy's skirmishers, whom we drove to their guns, which we unexpectedly found posted on our right, supported by a large force of cavalry. Their battery opened upon us, when we withdrew. In making this movement we were completely flanked by *another body* [italics added] of the enemy's cavalry, outnumbering my command at least six to one."[352]

Meanwhile, Capt. James H. Kidd recalled that "when within a mile of Hanover, the regiment turned off into a wheatfield, and mounting a crest beyond, came upon Fitzhugh Lee's Brigade, with a section of artillery in position, which opened upon the head of the regiment (then moving in column of fours), with shell wounding several men and horses."[353] (Note: It was Lee's Brigade that actually attacked the 6th Michigan. But it is likely that the brigade Kidd first sighted, which he believed to be Lee, was actually Chambliss. One critical aspect of Gray's report, which is supported by other 6th Michigan accounts, is the mention of two completely different large enemy forces. Clearly the second mentioned force is Lee's Brigade, the first must refer to Chambliss.)

These accounts have some important aspects in common that help locate approximately where this fighting occurred. In each case, it is clear that the encounter is a surprise; The Michigan men were not able to see the first large body of Confederates until after they left the road. The "crest," referred to by Kidd, was a prominent enough terrain feature to block the view of the opposing forces. This "crest" can only be what is now locally known as Mount Pleasant, a broad extent of higher ground, most of which is situated along, and south of, the Hanover-Littlestown Road, about two miles southwest of the center of Hanover. Any unit west of Mount Pleasant would remain out of sight from the Confederate artillery. Even today, this high ground dominates the view from those artillery positions. But if the 6th Michigan had remained on the Hanover-Littlestown Road after passing Mount Pleasant it would have been plainly visible to the Confederate gunners and Chambliss's Brigade (and vice-versa) while still on that road. They would not have been able to reach the Plum Creek crossing area without being seen by Confederates and coming under fire. But the accounts of the Michigan soldiers are consistent; they were not aware of the Confederate main body, and were not shelled by the Southern artillery, until after they moved well off the road and then crossed the high ground. They must have left the road while still under the cover of Mount Pleasant, and likely before they reached the Schwartz Schoolhouse.

Another aspect to consider is the evidence, or lack thereof, given by one local farmer. The Henry Sell residence was located where the Hanover-Littlestown Road crosses the Plum Creek. Historian George Prowell interviewed Sell; his eyewitness accounts were then used as a basis for a newspaper article in the *Hanover Herald* of Oct. 14, 1905.[354] Although he described a number of events that he witnessed that day, he did not mention any fighting on his land. If the 6th Michigan had been anywhere near Plum Creek when they were attacked, Sell's house and barn would have been completely enveloped in combat. (The southern portion of his land, however, was out of sight from his house; it is very likely that fighting took place in that area.)

The Kidd account mentions a wheat field where the regiment turned off the Hanover-Littlestown Road. It is uncertain exactly where this turnoff occurred, but at least two farmers in this area cited crop damage from Union cavalry moving through their property. Solomon and Samuel Schwartz each filed a federal damage claim. Both had wheat fields partially destroyed by Union cavalry passing over their land.[355] Determining exactly where the 6th Michigan column left the road is problematic, but a movement through the southeastern portion of Samuel's farm would have led the 6th directly over the crest of "Mount Pleasant" and into view of the Confederate gunners.

One tactical consideration that supports the above statements is the necessity of Confederate outposts to be positioned either along Narrow Drive or southwest of the Schwartz Schoolhouse. It would have been impossible for Southern patrols to protect the Confederate left flank had they been east of Mount Pleasant; their view to the south and west would have been blocked by the high ground. Control of Narrow Drive was also critical for the Confederates since that road led into the rear of their lines.

When the 6th Michigan was attacked it was in an area visible to the Confederate artillery, also to much of Chambliss's and Lee's Brigades. By that time, they had passed the Schwartz Schoolhouse, and the front of the column had likely reached the portion of Henry Sell's farm that was situated in Adams County. (Although Sell's house was in York County, much of his land was in Adams.) Whether any of the 6th reached York County before they were attacked is not certain, but it seems unlikely. Samuel Keller, who owned land to the east of the Sell property, made out a very detailed federal damage claim. His family listed several items taken by Confederates and damages done by Union Fifth Corps troops who encamped there the next day. Yet none of the family mentioned any damages done by actual fighting.[356] Also, it is hard to imagine the 6th could have reached Samuel's land, which at that time was controlled by a large portion of Chambliss's Brigade (the "large force" mentioned in Colonel Gray's report). When the 6th Michigan came into view of the Confederate main body, it was still several hundred yards from the Samuel Keller farm. Chambliss would have reacted immediately to this threat to his left flank. (An Adams County man named Jesse Keller owned large tracts of land between Mount Pleasant and McSherrystown. A small piece of his property was south of the Hanover-Littlestown Road in the Mount Pleasant area. Jesse filed a federal damage claim but he cited only damages which occurred from Ayres' division which encamped on his property on July 1, not from any combat on June 30.)

One damage claim that needs to be considered is that of John Shaeffer, whose land was south of the Sell property. According to Shaeffer and his son Edward, Confederate troops controlled their land throughout the skirmishing.[357] Their testimony would seem to indicate that the 6th Michigan crossed Mount Pleasant north of their farm.

Another consistent factor in the accounts of Union men is that the attack by Fitz Lee was also a surprise, to the extent that the 6th Michigan was almost completely surrounded. The regiment was in an area where Lee was able to move unseen upon their flank; only certain terrain features would have allowed this scenario to occur. It is impossible to know exactly where the Virginians launched their attack. But considering all the accounts of soldiers and civilians and examining the topography in this area, it appears that the 1st Virginia must have reached the high ground north of the John Shaeffer farm before they charged. By approaching this portion of "Mount Pleasant" from the south, Lee's Brigade would have come unseen upon the flank of the Michigan column.

As of this writing, this vicinity contains the only undeveloped parcels of land in the entire Hanover area where significant mounted combat took place.

APPENDIX J
GITT'S MILL SKIRMISHING

The accounts of 2nd Lt. Henry Potter (18th Pennsylvania) mention a small detail commanded by Capt. Thaddeus Freeland.[358] This patrol scouted to the right of the Hanover-Littlestown Road as the

Union column moved to Hanover. Freeland's detachment has been associated with the skirmishing at Gitt's Mill by several writers; the typical interpretation is that shots were fired there before the major fighting in Hanover. This narration can be traced directly to the writings of George Prowell from his *History of York County*.[359] That version has been repeated so often, a virtual "interpretations lockout" has unintentionally been put into place.

Freeland's party likely did make contact with the enemy somewhere south of the Hanover-Littlestown Road. Accounts included in the 18th Pennsylvania regimental history state that skirmishing took place to the right of the main column as it moved to Hanover.[360] However, there does not seem to be any actual *primary* source evidence to link this particular officer directly to the Gitt property. Although he was taken prisoner that day, the location of his capture was not documented in the regimental history. Even *if* Freeland's detachment was near Gitt's Mill, it cannot even be stated with certainty how it arrived at that site. Narrow Drive, Lovers Drive, Shibert Road and Sheppard Road were all present in 1863, and any one of them could have been used by his detail to reach the Gitt property. Freeland's detachment (or other detachment(s) of the 18th Pennsylvania) could have skirmished in many areas south of the Hanover-Littlestown Road without being on the property of Jeremiah Gitt.

At least one Confederate was killed in that general vicinity. Prowell's *History of York County* states that this casualty took place near Gitt's Mill, and was the first to occur on June 30. Prowell may have indeed been correct, but the basis of this assertion is unclear. One known reference to the incident was printed in local newspapers in 1903. But according to that article, William Gitt stated that the Southern soldier was killed not at the mill, but on Conewago Hill. (See also endnote 232 of chapter eight.)

(One troubling aspect of the standard interpretation concerns the element of surprise at the start of the battle. If the Gitt's Mill shooting occurred before the action in the town, Lieutenant Potter's rear guard would have still been well south of Hanover on the Littlestown Road. It is likely that any shooting near that site would have been heard by these cavalrymen, which would have alerted them to the presence of the enemy.)

It is possible that the typical interpretation is correct; shots may have taken place on Gitt's property before the major action occurred at Hanover. But the most intense activity, by far, that took place at Gitt's Mill occurred *after* the fighting in the town. Many of the specifics will likely never be known. But a combination of various civilian and military accounts provide a general idea of the troop movements and combat in the area.

Some of the most important evidence comes from property owners such as Jeremiah Gitt. In his federal damage claim, Gitt and Casper Krepps (one of his workers) specifically mention Michigan troops fighting on his land.[361] Since the 1st and 7th Michigan regiments were well north of Hanover before any Confederates were even close to the mill, these eyewitness statements must refer to 5th or 6th Michigan men who reached this area after the "main" battle had already started. (Note that any difficulties with the standard interpretation are resolved by the testimony given by Gitt and Krepps.)

Possibly the mill was occupied after Lee's Brigade struck the 6th Michigan. The main body of the 6th did not move through Gitt's property; the regiment encountered Confederate forces north of there and retreated away from that vicinity. The exception was the men of Companies B and F under Captain Weber, who had been ordered to protect the rear of the regiment. Weber's detachment was cut off well behind the Confederate lines for several hours. Some of that detail may have taken shelter in the mill for a time. However, of the several 6th Michigan accounts I have been able to locate, none refer to a mill site.

I believe it is more likely that men of the 5th Michigan occupied the mill. The account of John Bigelow, and the federal damage claim of Emanuel Wildasin, indicate that some of the 5th were deployed south of the Hanover-Littlestown Road as the regiment moved toward Hanover.[362] (The private lane known as Sheppard Road was a wartime road and ran parallel with the Hanover-Littlestown Road, directly to Gitt's Mill. Another critical terrain feature was the ridge along which Lovers Drive runs, which overlooks the Hanover-Littlestown Road. Safe passage to Hanover required at the very least scouting parties, if not a heavy skirmish force, to patrol "Lovers Drive Ridge." Note: At least a portion of Lovers Drive was the boundary between the lands of Gitt and Samuel Schwartz.) By the time elements of the 5th Michigan reached Gitt's property, much of Lee's Brigade had fallen back to the north and east of that site. This movement may have allowed skirmishers to take position in the mill for a time.

Michigan troops likely did not hold the mill for long. Any Union forces there were in danger of being cut off from their own supports, if that situation had not already occurred. A few hundred yards to the north, Confederate troops held higher ground than the mill site itself. But for a time, Michigan troops on the Gitt property and that of Samuel Schwartz, likely exchanged a lively fire with Southern skirmishers on the Shaeffer farm and along Narrow Drive.

The roads through Jeremiah Gitt's property saw significant troop movements, and his land saw extensive skirmishing. Concerning logistics, terrain, and tactics, the names Gitt's Mill, Conewago Hill, and Schwartz Schoolhouse should be regarded as essential parts of the events on June 30, 1863.

APPENDIX K
WHERE WERE THE WAGONS PARKED?

(In the last few years, various residents south of Hanover have had to contend with relic hunters on their property, some of whom did not receive permission to be on the land. At least one individual has even disturbed land that is posted as private property. Worse yet, it is doubtful that any of these finds have been donated to any research or historical facility. The only ones who have benefited are the trespassers. One relic hunter boasted to a landowner that after discovering a number of bullets, he "knew" the location where the captured wagons had been parked. But according to Capt. W. W. Blackford, one of Stuart's aides, and Col. Richard Beale of the 9th Virginia, every captured wagon was filled with provisions.[363] Most were laden with oats; others contained bread, hardtack, whiskey, sugar, hams, etc. None contained bullets or munitions. Unless the oats in those days were grown with some rather unique qualities, what this individual had likely found was the ammunition from one of the Confederates' own wagons, which would have been located in several areas south of town behind the Southern lines.)

A few local post war historians have made statements as to where the captured wagons were located when they were in park. In his 1907 *History of York County, Pennsylvania*, George Prowell wrote that "the wagon train was parked about two miles southwest of Hanover," but did not refer to a specific farm site.[364] By that time, even though Prowell had conducted a number of interviews with local citizens regarding the battle, he apparently had not been able to find a firsthand witness as to the wagon train location. In 1945, Hanoverian Robert E. Spangler was quoted as saying that the wagon train was parked "between the Westminster and Becker Mill Roads on lands of Samuel Keller and Henry Gotwalt…"[365] Spangler's statement, which has sometimes been misconstrued as an actual eyewitness account, is another facet of the battle that has been reprinted so many times that it has become accepted as fact. (Spangler was not born until 1867, and he did not cite any primary source for his information.)

If Spangler was referring to land owned by Gotwalt *after* 1863, his statement may be valid. But if this sentence refers to land actually owned by those men during the battle, then several problems exist with his statement.

Almost all of Keller's property was actually located west of the Westminster Road, which contradicts the Spangler statement. (The only exception was a very thin slice of land between the road and the Jacob Forney property.) Worse yet, the great majority of his 161 acres was in range of Union artillery on Bunker Hill. It seems inconceivable that Stuart would have placed his prized captured wagons in such a vulnerable position. No accounts have surfaced at this time of any Union gunners or officers sighting large numbers of wagons, which would have been an irresistible target. Meanwhile, the southern portion of Keller's land was wooded and would have created difficulties for parking a large number of wagons. Evidence given in Samuel Keller's federal damage claim only adds to the difficulties in Spangler's interpretation. No family member made any mention of damages occurring from Confederate wagons, even as great pains were taken to document other losses.[366]

Another difficulty with the Spangler statement is that in 1863, the Keller and Gotwalt lands did not share a common boundary. In fact, between their two parcels was a thirty-eight acre tract owned by John Christian Graby.[367] If the wagons were parked on both the Keller and Gotwalt property, the train must have been divided, unless Spangler is referring to land owned by Gotwalt *after* 1863.

Henry Gotwalt's home was about half a mile south of Keller's. The Gotwalt farm does seem to be a likely area for the wagons. This location was safely behind the battle lines, and shielded from the view of Union troops by the woods on the southern portion of the Keller farm. But here again, Spangler's statement regarding that location has introduced some misconceptions. Tax and deed records indicate that Gotwalt owned fifteen acres of land in 1863, but that also was on the west side of the Westminster Road (not between there and the Beck Mill Road). Not until at least 1864 did Gotwalt begin to acquire any property east of the Westminster Road.[368] In 1863, the wagons certainly could have been parked on both sides of the Westminster Road, but only those west of the road would have actually been on Henry Gotwalt's property.

One intriguing piece of evidence is found in the state damage claim of John Sheaffer. Sheaffer owned fifty acres in Adams County, and an adjoining forty acres in York County, and his property was well behind the Confederate lines. John and his son Edward both stated that "Rebel forces encamped" on land "adjacent to" the Sheaffer premises.[369] Their use of the term "encamped" would seem to indicate the wagon train and locates the site west of the Westminster Road. Henry Gotwalt was one of several landowners whose property bordered Shaeffer's land.

Another logistical concern was important at that point. By the time the wagons arrived, Stuart would have been aware that a movement through Hanover was no longer possible. The placement of the wagons would likely have been determined by this concern. According to the 1860 York County map, the only road from this area leading directly to the Baltimore Pike was Fairview Road.[370] Although some of the Confederates' own supply wagons were located closer to Hanover, the captured Federal vehicles were most likely placed well south of the Keller residence, possibly even south of the Gotwalt residence. In this way, the wagons would have been well out of sight and range of Union artillery, while still near the Fairview Road withdrawal route. Ample water sources were available in this vicinity; a branch of Plum Creek, the south branch of the Conewago Creek, and Indian Run all flow through the general area. One local historian, Clark B. Wentz, wrote that when the wagons "reached the top of Conewago Hill they turned right on what is now Fairview Road," and this interpretation, suggesting that the wagons never made it past Conewago Hill, must also be considered.[371] With no improved road north of Conewago Hill leading to the Baltimore Pike, it is possible that the Wentz interpretation is correct, and Stuart ordered the wagon train to park on

Conewago Hill itself. That location would have avoided the necessity of climbing this steep grade more than once.

As of this time, I have not found any federal or state claim that specifically cites damages for large numbers of wagons parked on a claimant's land. Since the captured Federal wagons were all loaded with provisions, as stated by Captain Blackford and Colonel Beale, there would be no large numbers of metal objects detectable to give away their location. Although much has been passed down in local lore, the precise location of the wagon parking area remains a mystery.

BIBLIOGRAPHY

In the research for this book, I attempted to utilize primary sources whenever possible. Many of those primary sources have never before appeared in any published work. What appears below in the "Records and Archival Collections" listing are only the general types and groups of documents used. For the specific documents cited, see endnotes.

RECORDS AND ARCHIVAL COLLECTIONS

Extensive use was made of documents from the following locations, with particular emphasis on those record groups listed:

National Archives
    Federal Quarter Master Damage Claims, Record Group 92, Vol. 2, Book 214
    Compiled Service Records, Union
    Compiled Service Records, Confederate
    Pension files
    Regimental Books (descriptive books, order books, and morning reports) of the following cavalry regiments: 5th Michigan, 6th Michigan, 18th Pennsylvania, 5th New York.

Library of Congress

Pennsylvania State Archives
    Pennsylvania Border Raid (Damage) Claims, filed with the State Auditor-General. Record Group 2.69, Auditor-General's Office.

Pennsylvania State Library

Adams County Historical Society
    Census Records of Adams, York County
    Deed Ledger Books
    Tax assessment books of Conewago, Union Townships

York County Archives
    Tax assessment books of Heidelberg, Manheim, West Manheim Townships
    Deed Ledger Books

York County Heritage Trust

Pennsylvania Room, Hanover Public Library
    Newspaper articles from the following publications:
    The Hanover Spectator
    The Hanover Citizen
    The Hanover Herald
    Hanover Record Herald
    Hanover Evening Herald

Gettysburg National Military Park Library Files (manuscript, letter, and diary collections)

Gettysburg Licensed Battlefield Guide Library Files (manuscript, letter, and diary collections)

PERIODICAL ARTICLES

Alexander, Ted. "Gettysburg Cavalry Operations, June 27-July 3 1863". *Blue and Gray*, Oct. 1988.

Callihan, David. "Jeb Stuart's Fateful Ride". *Gettysburg*, Issue #24, Jan. 2001.

Coddington, Edwin. "Prelude to Gettysburg: The Confederates Plunder Pennsylvania". *Pennsylvania History*, Volume 30, Number 2, April 1963.

Coski, John. "Forgotten Warrior: General William Henry Fitzhugh Payne". *North and South*, Volume 2, No 7, September 1999.

Ide, Horace K. "The 1st Vermont Cavalry in the Gettysburg Campaign", edited by Dr. Elliott Hoffman. (From original manuscript written in 1862 by Horace K. Ide, Sgt., Company D, 1st Vt. Cavalry). *Gettysburg* Issue #14, Jan. 1996.

Parsons, Henry. "Gettysburg: The Campaign was a Chapter of Accidents". *The National Tribune*, Aug. 7, 1890.

Paul, E. A. "Operations Of Our Cavalry: The Michigan Cavalry Brigade". *The New York Times*, August 6, 1863.

Ryan, Thomas J. "Kilpatrick Bars Stuart's Route to Gettysburg". *Gettysburg*, Issue #27, 2002.

Powell, David. "Stuart's Ride: Lee, Stuart, and the Confederate Cavalry in the Gettysburg Campaign". *Gettysburg* Issue #20, Jan. 1999.

Unknown Author. "The Thirteenth Regiment of Virginia Cavalry in Gen. J. E. B. Stuart's Raid into Pennsylvania". *The Southern Bivouac* (1883) pages 203-208.

MAPS

Carroll County 1862: *Martenet's Map of Carroll County, Maryland*, from surveys by S.J. Martenet, drawn and published by Simon J. Martenet, Surveyor and Civil Engineer, Baltimore.
1984 reproduction by Noodle Dorsey Press, Manchester, Md.

Carroll County 1877: *The Illustrated Atlas of Carroll County, Maryland*. Reprint of the 1877 edition by the Historical Society of Carroll Company, 1993.

Adams County 1858: *Survey of G. M. Hopkins, Civil Engineer*. M.S. and E. Converse Publishers, Philadelphia.

Adams County, 1872: *Atlas of Adams County, 1872* from surveys of D.J. Lake, Civil Engineer. Published by I.W. Field and Company, Philadelphia. Reproduction by Planks Suburban Press, Inc. Camp Hill, PA 1994.

York County 1860: *Shearer's Map.* From surveys by D.J. Lake. W.O. Shearer and D.J. Lake Publishers, Philadelphia.

York County 1876: *Atlas of York County.* Pomerey, Whitman, and Company From surveys by Beach Nichols, Published by Pomerey, Whitman, and Company, Philadelphia.

BOOKS AND MANUSCRIPTS

Anthony, William, editor. *The Battle of Hanover: Compiled from the Writings of George Prowell and Others.* (York County, Pennsylvania) Tuesday, June 30, 1863. Hanover, Pa.: William Anthony, 1945.

Balfour, Daniel T. *13th Virginia Cavalry.* Lynchburg, Va.: H. E. Howard, Inc. 1986.

Bates, Samuel P. *History of Pennsylvania Volunteers, 1861-1865.* 5 Volumes. Harrisburg, Pa.: B. Singerly, 1871.

Beale, George W. *A Lieutenant of Cavalry in Lee's Army.* Boston: Gorham Press, 1918.

Beale, Richard L. T. *History of the 9th Virginia Cavalry.* Richmond: B. F. Johnson Publishing Company, 1899.

Beaudry, Richard E., editor. *War Journal of Louis N. Beaudry, Fifth New York Cavalry.* Jefferson, N. C.: McFarland & Company, Inc., 1996.

Benedict, G. G. *Vermont in the Civil War, 1861-1865.* Two volumes. Burlington Vt.: Free Press Association, 1888. Reprinted from Salem, Mass.: Higginson Book Company

Beyer, W. F. and Keydel, O. F., editors. *Deeds of Valor, From Records in the Archives of the United States Government: How American Heroes Won the Medal of Honor.* Two Volumes. Detroit, Mich.: The Perrien-Keydel Company, 1907.

Blackford, W. W. *War Years With Jeb Stuart.* New York: Charles Scribner's Sons, 1945.

Boudrye, Louis N. *Historic Records of the Fifth New York Cavalry, First Ira Harris Guard.* Albany: S. R. Gray, 1865.

Busey, John W. and David G. Martin. *Regimental Strengths and Losses at Gettysburg.* Highstown, N. J.: Longstreet House, 1994.

Bush, B. Conrad, compiler. *Articles from Wyoming County Newspapers and Letters from Soldiers of the 5th New York Cavalry.* West Falls, N. Y.: Bush Research, 2000.

Carter, Lt. Col. William R. *Sabres, Saddles, and Spurs.* Edited by Walbrook D. Swank, Colonel, USAF (Ret.). Shippensburg, Pa. Burd Street Press publication, printed by Beidel Printing House, 1998.

Coddington, Edwin B. *The Gettysburg Campaign, A Study in Command.* New York: Charles Scribner's Sons, 1968.

Cooke, John Esten. *Wearing Of The Gray*: Being Personal Portraits, Scenes, and Adventures of the War. Bloomington In.: Indiana University Press, 1959.

Downey, Fairfax. *A Clash of Cavalry: The Battle of Brandy Station.* New York: David McKay Company, 1959.

Driver, Robert J., Jr. *1st Virginia Cavalry.* Lynchburg, Va.: H. E. Howard, Inc., 1991.

Driver, Robert J., Jr. *5th Virginia Cavalry.* Lynchburg, Va.: H. E. Howard, Inc., 1997.

Driver, Robert J., Jr. *10th Virginia Cavalry.* Lynchburg, Va.: H. E. Howard, Inc., 1992.

Driver, Robert J., Jr. *13th Virginia Cavalry.* Lynchburg, Va.: H. E. Howard, Inc., 1992.

Driver, Robert J., Jr. and H. E. Howard. *2nd Virginia Cavalry.* Lynchburg, Va.: H. E. Howard, Inc., 1995.

Dyer, Frederick H. *A Compendium of the War of the Rebellion.* Three volumes. New York: Yoseloff, 1959.

Fishel, Edwin C. *The Secret War for the Union: The Untold Story of Military Intelligence in the Civil War.* Boston and New York: Houghton Mifflin Company, 1996.

Fox, William F., editor. *New York at Gettysburg.* Three volumes. Albany, N. Y.: J. B. Lyon Company, 1902.

Gettysburg Battle Field Commission of Michigan. *Michigan at Gettysburg, July 1st, 2nd, and 3rd, 1863. June 12, 1889. Proceedings Incident to the Dedication of the Michigan Monuments upon the Battlefield of Gettysburg, June 12, 1889.* Detroit: Winn & Hammond, 1889.

Gibson, John, Historical Editor. *History of York County, Pa.* Chicago: F. A. Battey Publishing Company, 1886.

Gillespie, Samuel L. ("Lovejoy"). *A History of Company A, First Ohio Cavalry, 1861-1865.* Washington Court House, Oh.: Ohio State Register Press, 1898.

Hackley, Woodford B. The Little Fork Rangers: *A Sketch of Company D, Fourth Virginia Cavalry.* Richmond, Va.: Dietz Printing Company, 1927.

Hanover Area Chamber of Commerce, Publisher. *Encounter at Hanover: Prelude to Gettysburg.*

Hanover, Pa.: Hanover Chamber of Commerce, 1962.

Harrell, Roger H. *The 2nd North Carolina Cavalry.* Jefferson, N. C. and London: McFarland and Company, Inc.

Haven, Eloise Amy, compiler and editor. *In The Steps of a Wolverine: The Civil War Letters of a Michigan Cavalryman* (Allen D. Pease, Company B, 6th Michigan Cavalry). Kentwood, Mich.: self-published work, 2005.

Harris, Samuel. *The Personal Reminiscences of Samuel Harris.* Detroit, Mich.: The Robinson Press, 1897.

Hudgins, Garland C. and Kleese, Richard B., eds. *Recollections of an Old Dominion Dragoon: The Civil War Experiences of Sgt. Robert S. Hudgins II, Company B, 3rd Virginia Cavalry.* Orange, Va.: Publisher's Press, Inc. 1993.

Isham, Asa B. *An Historical Sketch of the Seventh Regiment Michigan Volunteer Cavalry from Its Organization, in 1862, to Its Muster Out, in 1865.* New York: Town Topics Publishing, 1893.

Johnson, Robert U., and Clarence C. Buel, editors. *Battles and Leaders of the Civil War, Being for the Most Part Contributions by Union and Confederate Officers*, Based Upon "The Century War Series." Four volumes. New York: Century Company, 1884-1889.

Kidd, James H. *Personal Recollections of a Cavalryman: With Custer's Michigan Cavalry Brigade in the Civil War.* Ionia, Mich.: Sentinel Printing, 1908. Reprinted from Grand Rapids, Mich.: The Black Letter Press, 1969.

Klein, Frederic Shriver, editor, with the collaboration of W. Harold Redcay and G. Thomas LeGore. *Just South of Gettysburg: Carroll County, Maryland in the Civil War-Personal Accounts and Descriptions of a Maryland Border County, 1861-1865.* Westminster, Md.: The Newman Press, 1963.

Krick, Robert K. *9th Virginia Cavalry.* Lynchburg, Va.: H. E. Howard, Inc., 1982.

Krick, Robert K. *Lee's Colonels: A Biographical Register of the Field Officers of the Army of Northern Virginia.* Dayton, Oh.: Morningside House, Inc., 1992.

Ladd, David L., and Audrey J., editors. *The Bachelder Papers: Gettysburg in Their Own Words.* Three volumes. Dayton, Oh.: Morningside House, Inc., 1994.

Longacre, Edward G. *The Cavalry at Gettysburg: A Tactical Study of Mounted Operations during the Civil War's Pivotal Campaign 9 June-14 July 1863.* Cranbury, N. J.: Associated University Press, 1986.

Longacre, Edward G. *Custer and His Wolverines: The Michigan Cavalry Brigade 1861-1865.* Conshohocken, Pa.: Combined Publishing, 1997.

Manarin, Louis H., Compiler. *North Carolina Troops 1861-1865, A Roster*. Raleigh, N. C.: The North Carolina State University Print Shop, 1968.

McClellan, Henry B. *The Life and Campaigns of Major General J. E. B. Stuart*. Boston: Houghton Mifflin & Company, 1885.

Moore, Robert H., II. *The 1st and 2nd Stuart Horse Artillery*. Lynchburg, Va.: H. E. Howard, Inc., 1985.

Morgan, James A., III. *Always Ready, Always Willing: A History of Battery M, Second United States Artillery from It's Organization Through the Civil War*. Gaithersburg, Md.: Olde Soldier Books, Inc., no date.

Mosby, John S. *The Memoirs of Colonel John S. Mosby*, edited by Charles Wells Russell, 1917. Bloomington, In: Indiana University Press, 1959.

Murray, R. L., editor. *Letters from Gettysburg: New York Soldiers' Correspondence from the Battlefield*. Wolcott, N. Y.: Benedum Books, 2005.

Nanzig, Thomas P. *3rd Virginia Cavalry*. Lynchburg, Va.: H. E. Howard, Inc., 1989.

Nesbitt, Mark. *Saber and Scapegoat: J. E. B. Stuart and the Gettysburg Controversy*. Mechanicsburg, Pa.: Stackpole Books, 1994.

New York Monuments Commission for the Battlefields of Gettysburg and Chattanooga: *Final Report on the Battlefield of Gettysburg*. Three Volumes. Albany, N. Y.: J. B. Lyon Company, Printers, 1902.

Nicholson, John P. *Pennsylvania at Gettysburg: Ceremonies at the Dedication of the Monuments Erected by the Commonwealth of Pennsylvania to Major General George G. Meade, Major General Winfield S. Hancock, Major General John F. Reynolds, and to Mark the Positions of the Pennsylvania Commands Engaged in the Battle*. Two volumes. Harrisburg, Pa.: Wm. Stanley Ray, State Printers, 1893.

Phipps, Michael. *Come On You Wolverines: Custer at Gettysburg*. Gettysburg, Pa.: Farnsworth House Military Impressions, 1996.

Phisterer, Frederick. *New York in the War of the Rebellion, 1861-1865*. Albany, N. Y.: J. B. Lyon Company, 1912.

Prowell, George R. *History of York County, Pennsylvania*. Chicago: J. H. Beers and Company, 1907

Publication Committee of the Regimental Association. *History of the Eighteenth Regiment of Cavalry, Pennsylvania Volunteers (163rd Regiment of the Line) 1862-1865*. New York: Wynkoop, Hallenbeck, Crawford, 1909.

Raus, Edmund J., Jr. *A Generation on the March: The Union Army at Gettysburg*. Lynchburg, Va.:

H. E. Howard, Inc., 1987.

Reily, John T. (Publisher). *History and Directory of the Boroughs of Adams County.* Gettysburg, Pa.: J. E. Wible, Printer, 1880.

Robertson, James I., Editor in Chief. *Southern Historical Society Papers, 1876-1959.* Wilmington, N. C.: Broadfoot Publishing Company, 1992.

Robertson, Jno., compiler. *Michigan in the War.* Lansing, Mich.: W. S. George & Company, 1880.

Rummel, George A., III. *Cavalry on the Roads to Gettysburg: Kilpatrick at Hanover and Hunterstown.* Shippensburg, Pa.: White Mane Publishing Company, Inc. 2000.

Stiles, Kenneth L. *4th Virginia Cavalry.* Lynchburg, Va.: H. E. Howard, Inc., 1985.

Stowe, Mark S. *Company B: 6th Michigan Cavalry.* Grand Rapids, Mich.: Mark S. Stowe Publishing, 2002.

Trout, Robert J. Galloping Thunder: The Story of the Stuart Horse Artillery Battalion. Mechanicsburg, Pa.: Stackpole Books, 2002.

Trout, Robert J. *They Followed the Plume: The Story of J. E. B. Stuart and His Staff.* Mechanicsburg, Pa: Stackpole Books, 1993.

United States War Department. *The War of the Rebellion: A Compilation of the Official Records of the Union and Confederate Armies.* Volume 27, Parts 1-3. Washington, D. C.: Government Printing Office, 1889.

United States War Department. *Supplement to the Official Records of the Union and Confederate Armies.* Addendum to Series I, Volume 27, Parts 1-3. Wilmington, N. C.: Broadfoot Publishing Company, 1995.

Utley, Robert M. *Frontier Regulars: The United States Army and The Indian, 1866-1891.* New York: Macmillan Publishing Company, Inc. and London: Collier Macmillan Publishers, 1973.

Warner, Ezra J. *Generals in Blue: Lives of the Union Commanders.* Baton Rouge: Louisiana State University Press, 1964.

Warner, Ezra J. *Generals in Gray: Lives of the Confederate Commanders.* Baton Rouge: Louisiana State University Press, 1959.

Wentz, Clark B. *History of West Manheim Township.* Unpublished manuscript, 1969. Copy on file in Pennsylvania Room, Hanover Public Library.

Wittenberg, Eric J. and Petruzzi, J. David. *Plenty of Blame to Go Around: Jeb Stuart's Controversial Ride to Gettysburg.* New York: Savas Beatie, 2006.

ENDNOTES

PROLOGUE

[1] 1860 Federal Census for Borough of Hanover.

[2] Almost all York and Adams County roads at that time were dirt, although their condition varied greatly. If a major route was administered by a turnpike organization, tolls were charged and then used to perform routine maintenance. A few of these "pikes," but not necessarily all, were macadamized with a surface of crushed stone and gravel. Most, however, were not under any official maintenance until after the Civil War. At least two exceptions can be made. By 1860, much of the road leading toward Baltimore was under the oversight of the Hanover and Maryland Line Turnpike Company. Also, the route between Hanover and Abbottstown was under the direction of the Hanover and Berlin Turnpike Company at the time.

[3] Article entitled "The Public Commons," *The Hanover Herald,* March 31, 1906.

[4] Secondary account of Robert Spangler. From: William Anthony, editor, *The Battle of Hanover:* Compiled from the writings of George Prowell and others (Hanover, Pa.: William Anthony - editor, printer, and publisher, 1945), 144. Hereafter referred to as Anthony, *The Battle of Hanover.*

[5] *Ibid.*

[6] United States War Department, *War of the Rebellion: A Compilation of the Official Records of the Union and Confederate Armies,* 128 Vols. (Washington D. C.: Government Printing Office 1880-1901) Series I, Vol. 27, Part 2, 464-465. Hereafter cited as *O.R.* with reference to Series I.

[7] *Ibid.,* 465. Troop strength of the 35th Virginia Battalion from John Busey and David G. Martin, *Regimental Strengths and Losses at Gettysburg* (Highstown, N. J.: Longstreet House, 1994), 198. Hereafter referred to as Busey and Martin, *Regimental Strengths and Losses at Gettysburg.*

[8] Testimony of John Rife and Daniel Geiselman, November 10, 1868. From State Damage Claims of John Rife and Daniel Geiselman. See "Border Claims" filed with Pennsylvania State Auditor General, Record Group 2.69, Auditor General's Office. Microfilm records at the Pennsylvania State Archives or Adams County Historical Society. All Pennsylvania Auditor General's Claims hereafter referred to as State Damage Claims.

[9] Testimony of Alfred Burkee, Conrad Cramer, and Michael Reily, October 22, 1868. State Damage Claim of Michael Reily and Vincent Sneeringer.

[10] Letter of Daniel Trone to A. W. Eichelberger, June 30, 1863. Original letter was presented by Mrs. William Himes to the Hanover (now Guthrie) Public Library. It is now in the care of the library's Pennsylvania Room and is cited here with their kind permission. It was also reprinted (with a few minor changes of punctuation and to one sentence) in the work compiled by the Publication Committee of the Hanover Chamber of Commerce, *Encounter at Hanover: Prelude to Gettysburg* (Hanover, Pa.: Hanover Chamber of Commerce, publishers, 1962. Third printing from Shippensburg, Pa.: White Mane Publishing Company, Inc., 1988), 117. Hereafter referred to as *Encounter at Hanover.*

[11] *Ibid.*

[12] Article entitled "Civil War Days Here," *Record Herald,* January 31, 1922. Leib did not actually say that the Confederates shot *at* them, although his account seems to imply that scenario. He stated that he was able to escape because he was out of the range of their carbines. Possibly the Confederates were firing warning shots.

[33] It is not certain what path Stuart would have taken on the morning of June 30 had he not learned of the Union presence at Littlestown. His official report was not specific on that point. Possibly he would have moved to Hanover anyway. Most historians have asserted that Stuart's original intention was to head toward Gettysburg that morning. By going to Gettysburg he would probably discover information on the movements of Confederate infantry, possibly along the York Road. If that was his plan, it became a moot point once Stuart learned that Union cavalry had reached Littlestown.

CHAPTER TWO

[34] The directive from army headquarters that relieved Stahel of his position in the Army of the Potomac (Special Orders 174) was possibly the last order issued by Hooker. The instructions that placed Kilpatrick in charge of the Third Division was part of Special Order 98 from cavalry corps headquarters, and was possibly the first order issued by Pleasonton as a corps commander. See *O.R. Vol. 27, Part 3*, 373, 376.

[35] Ezra J. Warner, *Generals in Blue: Lives of the Union Commanders* (Baton Rouge, La.: Louisiana State University Press, 1964), 266.

[36] *Ibid.*, 149.

[37] *Ibid.*, 109.

[38] *O.R., Vol. 27, Part 1*, 991. In *Regimental Strengths and Losses at Gettysburg*, authors Busey and Martin posit a June 30 strength of 2,189 for Farnsworth and 2,345 for Custer. These figures suggest that the Third Division had about 4,534 soldiers at Hanover, not including the artillery of Elder and Pennington or the headquarters guard of Kilpatrick.

[39] In 1861, Congress passed legislation that new Union cavalry regiments would consist of twelve companies. Union volunteer infantry regiments typically retained the traditional ten company structure.

[40] Busey and Martin, *Regimental Strengths and Losses at Gettysburg*, 107.

[41] *Ibid.*, 206.

[42] *Ibid.*, 109.

[43] *Ibid.*, 108.

[44] Apparently, the entire 5th Michigan had Spencer rifles by that time. According to their Ordnance Returns, nine companies of the regiment reported having those weapons for the quarter ending June 30, 1863. No report exists for Companies D, E, and M for that quarter, but Company M did report having Spencers the previous quarter ending March 31.

Meanwhile, Companies D and E cited having Spencers in reports filed in other quarters later that year. Colonel Alger later confirmed that information. In a report given to the adjutant general of the army on July 1, 1880, he stated that by June 30, 1863, his regiment was armed with the Spencer rifle. Concerning the 6th Michigan, two companies, I and M, were not present with the regiment on June 30. Of the other ten, the following can be stated: Companies A, D, E, and H reported having Spencer rifles by that time, in returns for either the first or second quarter of 1863. Meanwhile, four companies reported having Burnside carbines by that time. For two other companies, no ordnance report exists for either the first or second quarter of that year. These statistics were taken from National Archives Record Group 156, *Records of Office of Chief of Ordnance - Summary Statements of Quarterly Returns of Ordnance*. Microfilm Series M1281.

[45] Busey and Martin, *Regimental Strengths and Losses at Gettysburg*, 109.

[46] Jno. Robertson (compiler), *Michigan in the War* (Lansing, Mich.: W. S. George and Company, 1880) 573, 574. Hereafter referred to as *Michigan in the War*. See also *O.R. Vol. 27, Part 3*, 350.

[47] June 29, 1863 diary entry of Dexter Macomber, 1st Michigan Cavalry. Clarke Historical Library, Central Michigan University. Transcript on file in Gettysburg Licensed Battlefield Guide Library, 1st Michigan Cavalry file. Hereafter referred to as Dexter Macomber diary.

[48] Kilpatrick's official report states that Custer's entire brigade reached Littlestown by 10 P. M., June 29, but there is substantial evidence that the 5th and 6th Michigan did not reach the area until the early morning of June 30. The accounts of James H. Kidd, William Baird, Daniel Stewart (all of the 6th Michigan), and John Allen Bigelow (5th Michigan) state that these two regiments reached Littlestown on the morning of June 30. See speech of James H. Kidd in: Monument Commission, *Michigan at Gettysburg, July 1st, 2nd, and 3rd, 1863. June 12, 1889 – Proceedings Incident to the Dedication of the Michigan Monuments at Gettysburg, June 12, 1889* (Detroit, Mich.: Winn and Hammond, Printers and Binders, 1889), 138. Hereafter referred to as *Michigan at Gettysburg*. Copies of accounts of Baird, Stewart, and Bigelow from Gettysburg National Military Park Library (all cited below).

[49] Reverend Louis N. Boudrye (Chaplain), *Records of the Fifth New York Cavalry* (Albany, N. Y.: S. R. Gray, 1865), 64. Hereafter referred to as Boudrye, Records of the Fifth New York Cavalry.

[50] Diary of Charles Blinn, Company A, 1st Vermont. Copy in file of Gettysburg National Military Park Library. Hereafter referred to as Charles Blinn diary.

[51] Samuel L. Gillespie (pen name Lovejoy), *A History, Company A, First Ohio Cavalry, 1861-1865* (Washington Court House, Oh.: Ohio State Register Press, 1898), 147. Hereafter referred to as *Samuel Gillespie*. In the official Order of Battle for the Gettysburg campaign, Company A of the 1st Ohio is listed as being the headquarters guard for Gen. David McMurtrie Gregg, commander of the Second Cavalry Division, Army of the Potomac. But, according to Gillespie, Company A performed that service, at least for a time, for Kilpatrick. In one recent cavalry study, author George A. Rummel III cites overwhelming evidence that Gillespie is correct. At that time, Company A (along with Company C) was Kilpatrick's escort. See George A. Rummel III, *Cavalry On the Roads To Gettysburg: Kilpatrick at Hanover and Hunterstown*, (Shippensburg, Pa.: White Mane Publishing Co., Inc., 2000), 141, note 4.

[52] *Samuel Gillespie*, 149. On the 1858 Adams County map, two different names are cited for the same town. The county-wide portion lists the town's name as Berlin, while the community close-up section lists the town by its current name, East Berlin.

CHAPTER THREE

[53] Letter from Miss Sallie C. Shriver to her sister, Mrs. Elizabeth J. Myer. Sallie stated that it was written at "12 o' clock Tuesday night" (June 30), although some of the letter was finished a few days later. According to Sallie, the first Confederates reached the Shriver property at 10 P. M. followed "in a short time" by Gen. Fitzhugh Lee and many others. Sallie's letter is hereafter referred to as S. C. Shriver letter. See Frederic Shriver Klein, editor, with the collaboration of W. Thomas Redcay and G. Thomas LeGore. *Just South of Gettysburg: Carroll County, Maryland in the Civil War; Personal Accounts and Descriptions of a Maryland Border County, 1861-1865* (Westminster, Md.: The Newman Press, 1963), 186. Hereafter referred to as *Just South of Gettysburg*.

[54] *Ibid.*, 187. In another letter, dated July 4, 1863, F. A. Shriver wrote that "General Stuart had his headquarters at Westminster that night while Lee had his up here..." *Ibid*, 184. Hereafter referred to as F. A. Shriver letter.

[55] *O.R.*, Vol. 27, Part 2, 695, 696.

[56] Beale, *Ninth Virginia Cavalry*, 78.

[57] Busey and Martin, *Regimental Strengths and Losses at Gettysburg*, 197. The 15th Virginia was also in this brigade but was on detached duty and not present during Stuart's expedition.

[58] Beale, *Ninth Virginia Cavalry*, 81. Also see *Supplement to the Official Records of the Union and Confederate Armies, Record of Events - Volume 69, Virginia Troops (Confederate) - Cavalry*, report of Company H, 9th Va., (Wilmington, N. C.: Broadfoot Publishing Company, 1995), 843. Hereafter referred to as *Supplemental Records*.

[59] According to Col. Richard Beale, the 13th Virginia was at the front of the column and was directly in front of the 9th Virginia. Although he does not explicitly state the positions of the 2nd North Carolina or 10th Virginia, his account seems to indicate that the 2nd North Carolina was nearby when the first shots were fired. He does not mention the 10th Virginia being in the area at that time. The 2nd North Carolina was heavily involved in the fighting, while the 10th Virginia was not engaged in the major attacks. These facts, along with Beale's statements, indicate that the 10th was likely the last regiment in Chambliss's Brigade to reach Hanover. See Beale, *Ninth Virginia Cavalry*, 82.

[60] Most historians have assumed that Herbert Shriver accompanied Stuart personally that morning. This explanation was held by William, one of Herbert's sons. William, however, did not witness the events (he was not born until after 1863). But another family member wrote that Herbert's father (also named William) claimed that the boy was "under the immediate protection of Fitz Lee." It is possible that guides accompanied both columns. But the road taken by Stuart's main body was a direct path to Hanover, while the route taken by Lee's men required a thorough knowledge of several secondary roads through a very isolated area. William's account, although it was written many years later, seems to indicate that Herbert accompanied Fitzhugh Lee. His letter was based on his father's description of incidents that occurred during his ride. The context of this letter suggests very strongly that those events took place along the Baltimore Pike, very near the Pennsylvania-Maryland line, on the route taken by Lee's Brigade. See *Just South of Gettysburg*, 181, 184, 187, 200-204, for the following accounts: S. C. Shriver letter; F. A. Shriver letter; Louis Shriver account which was published in *The Westminster Times* of March 26, 1937; and letter of William H. Shriver, (son of Herbert).

[61] John Esten Cooke, *Wearing of the Gray: Being Personal Portraits, Scenes, and Adventures of the War* (Bloomington Indiana Press, 1959), 239. This work was initially published in 1867, being composed of previously written articles. Hereafter referred to as John Esten Cooke, *Wearing of the Gray*.

[62] Testimony of Adam Leese, William Leese, and Amos Feeser, November 4 and 7, 1868, and July 9, 1883. Federal Damage Claim of Adam Leese, National Archives Record Group 92, Vol. 2, Book 214, Claim 737. The seven other Union Township, Adams County residents referred to in the text actually filed Pennsylvania state damage claims. See State Damage Claims of John Baker, John Baublitz, Elizabeth Feeser, George Feeser, David Gobrecht, Frederick Lohr, and William Sterner.

[63] Beale, *Ninth Virginia Cavalry*, 81.

[64] Testimony of John Baublitz, October 23, 1868. State Damage Claim of John Baublitz.

[65] Testimony of Edmund and Catherine Lippy, November 10, 1868, and June 12, 1882. Federal

Damage Claim of Josiah Gitt, National Archives Record Group 92, Vol. 2, Book 214, Claim 1546.

[66] Letter of Capt. William A. Graham, April 5, 1886. Hereafter referred to as Graham Letter. From *The Bachelder Papers – Gettysburg In Their Own Words, Vol. III* (Dayton, Oh.: Morningside House, 1994), 1373. Hereafter referred to as *The Bachelder Papers*. These letters were transcribed, edited, and annotated by David L. and Audrey J. Ladd from the Bachelder Papers in the New Hampshire Historical Society.

[67] Testimony of Jacob Leppo, Ephraim Nace, Jesse Houck, Benjamin Wentz, John Lohr, Perry Mathias, and David Baughman, October 23, 1868. From State Damage Claims of the same individuals. These claimants lived on/near the roads cited in the text. Parts of Grand Valley Road have changed course with the construction of a man-made lake, known locally as Long Arm Dam. As these roads branch off the route of Stuart's main column, it suggests that detachments were moving through here while the main column continued toward Hanover. Captain William Graham's statements and Ephraim Nace's interview with George Prowell (see footnote 69) also support this interpretation.

[68] Charles Francis Adams, Jr., *A Cycle of Adams Letters, Vol. 2* (Boston and New York: Houghton Mifflin & Company, 1920), 2, 3-6. Reprinted in Fairfax Downey, *A Clash of Cavalry: The Battle of Brandy Station, June 9, 1863* (New York: David McKay Company, Inc., 1959), 32-34.

[69] Statement of Ephraim Nace, from his interview with local historian, George Prowell. Nace's accounts formed the basis of an article written by Prowell entitled "Forty-One Years Ago." See *Hanover Record Herald*, June 30, 1904. Hereafter referred to as Ephraim Nace account.

[70] Jacob Leppo testimony, October 23, 1868. State Damage Claim of Jacob Leppo. Leppo lived along Impounding Dam Road (modern name) in West Manheim Township.

[71] Departure times given in Union accounts indicate that the leading troops began the ride to Hanover early that morning. Kilpatrick's campaign report states that the division marched at "daylight." The statement of D. H. Robbins (5th New York) corroborates the general's report. Robbins noted that Custer led off with his brigade at "daylight." Dexter Macomber (1st Michigan) was more specific. In his diary, he mentioned that his regiment saddled "at 4 a. m." Lieutenant Pennington, commanding Battery M, 2nd U. S. Artillery, stated that his battery left camp "at 5 a. m." In Farnsworth's brigade, Charles Blinn (1st Vermont) stated that he left Littlestown "at daybreak." William Porter Wilkin (alexander) noted that he began the march at "5 o'clock A. M." One account that is somewhat at odds with the others is that of Lt. Henry C. Potter (18th Pennsylvania) who stated that the division "left Littlestown at 8 o'clock A. M.," and since the 18th "was the last to leave... I did not leave there until 10 A. M." Although Potter commanded the rear guard detail, and would have started somewhat later than the leading troops, his times given seem somewhat off. For the previous statements, see the following sources: Kilpatrick's time taken from *O.R. Vol. 27, Part 1*, 992. Robbins' statement from *The National Tribune*, May 20, 1915. Blinn and Macomber statements taken from diaries cited in earlier footnotes. Pennington quote from *Supplemental Records, Vol. 5,* 285 (Addendum to the Reports of Vol. 27). Original Wilkin diary is in possession of Richard K. Irish, Marshall, Virginia. Copies or transcripts of these diaries are on file in the Gettysburg National Military Park Library. Potter account taken from Publication Committee of the Regimental Association, *History of the Eighteenth Regiment of Cavalry, Pennsylvania Volunteers (163rd Regiment of the Line) 1862-1865.* (New York: Wynkoop, Hallenbeck, Crawford, 1909), 87. Hereafter referred to as *Eighteenth Pennsylvania Regimental History*.

[72] Kilpatrick did not state where he rode in the column during its march to Hanover. It has been

assumed by most historians that he was at the front of the 1st Michigan, likely because civilian accounts placed him at the front of his troops. However, a large gap was present between the front of Farnsworth's brigade and the rear of the 7th Michigan that preceded them. If Kilpatrick had been riding at the head of Farnsworth's brigade, it would have appeared to any civilians that he was at the head of the column.

73 By 10:00 A. M., the gap between the 7th Michigan and the front of Farnsworth's brigade was significant since the 1st and 7th Michigan were near Abbottstown when the first shots were fired. The order of regiments in Farnsworth's brigade was taken from the report of Maj. John Hammond, excepting that evidence suggests the 1st West Virginia led Farnsworth's brigade that morning. See *O.R. Vol. 27, Part 1,* 1008. (See Appendix D for a discussion of the positions of the 1st West Virginia and 1st Vermont. Concerning the locations of the ambulance wagons, see endnotes 90 and 91.)

74 Reports of Colonel Alger and Colonel Gray. See *Michigan in the War,* 578, 580. Also *Supplemental Records Part 1 (Reports), Vol. 5,* 271, 263-264. The regimental reports submitted by officers of the 1st, 5th, 6th, and 7th Michigan were edited by Custer, who then submitted them under a consolidated brigade report. In the meantime, the individual regimental reports were reprinted in *The New York Times* of August 6, 1863. Later, they were included in *Michigan In the War* and the *Supplemental Records.*

75 According to the 1863 Conewago Township tax assessment book, the mill owner's name was spelled Duttera. This is one area where the Hanover-Littlestown Road has seen a major course change. The section of the road that passed the mill is still in existence, but approximately three-fourths mile of the original route is now a private drive. The modern road (Route 194) now crosses the creek approximately a few hundred yards south of the original mill site then rejoins the original trace several hundred yards later.

76 George Hoch statement from article entitled "Battle of Hanover," *The Hanover Herald,* October 1, 1910. Hereafter referred to as George Hoch account.

77 Samuel Forney narrative from article entitled "Forney Boys Experience," *The Hanover Herald,* July 8, 1905. Hereafter referred to as Forney article.

78 George Spangler account from article entitled "Reminiscences of the Civil War Times," *The Hanover Herald,* September 2, 1905. Hereafter referred to as George Spangler account. George was the older brother of local historian Robert Spangler, who gave a secondary account of this same incident. See Anthony, *The Battle of Hanover,* 143.

79 Accounts of Carrie Moul and Rebecca Scheurer. Excerpts of a journal written by Moul were reprinted in an article entitled "War-Time Impressions." See *The Hanover Herald,* November 9, 1914. Hereafter referred to as Carrie Moul account. Scheurer's account is from an article entitled "The Battle of Hanover," *The Hanover Herald,* June 18, 1910. Hereafter referred to as Rebecca Scheurer account.

80 George Spangler account.

81 Account of Rev. William K. Zieber, from article entitled "The Battle of Hanover," *The Hanover Herald,* July 15, 1905. Hereafter referred to as Reverend Zieber account.

82 Various sources differ as to the arrival time of the Union troops. It is not always clear if the accounts refer to the arrival of the 1st and 7th Michigan at the head of the column or the leading regiments of Farnsworth's brigade, which arrived later. Some state that Union cavalry arrived about 8:00 A. M. One was Carrie Moul, in her account cited above. *The Hanover Spectator* of July 3, 1863, also stated an 8:00 A. M. arrival time. Kilpatrick's report gives the time of arrival at 10:00 A. M. (There seems to be a puzzling discrepancy in the general's report. Kilpatrick also

stated that the march began "At daylight." His men had started the movement from the area just east of Littlestown; it would not have taken from daybreak until 10:00 A. M. to reach Hanover.) Some officers in Farnsworth's brigade gave even later times. Major Hammond stated that the 5th New York entered the town "About noon." According to Colonel Richmond (1st W. Va.) "the advance of the column arrived about noon". Even if the times given by Hammond and Richmond are well off the mark, there was clearly a significant gap between the 7th Michigan and the front of Farnsworth's brigade. For the statements of Kilpatrick, Richmond, and Hammond, see *O.R Vol. 27, Part 1*, 992, 1005, 1008.

[83] Letter of William Porter Wilkin to his wife dated July 21, 1863. Reprinted in the Ohio *Athens Messenger*, August 13, 1863. Transcript on file in Gettysburg National Military Park Library.

[84] Charles Blinn diary.

[85] Account of Sgt. Horace K. Ide, Company D, 1st Vermont. From article titled "The 1st Vermont Cavalry in the Gettysburg Campaign," *Gettysburg Magazine*, issue Number Fourteen, 12. Taken from original manuscript written by Ide in 1872. Hereafter referred to as Horace Ide account.

[86] Account of Capt. Henry C. Parsons, 1st Vermont. From article entitled "Gettysburg: The Campaign Was a Chapter of Accidents." *The National Tribune*, August 7, 1890. Hereafter referred to as Parsons, *National Tribune* article.

[87] Several sources verify that the head of the division had reached the Abbottstown area before the first shots were fired at Hanover. One was Dexter Macomber of the 1st Michigan. Another was a local boy, Josiah Thoman (cited in endnote 165). Also see Asa B. Isham, *An Historical Sketch of the Seventh Regiment Michigan Volunteer Cavalry from its Organization, in 1862, to its Muster Out, in 1865 (*New York: Town Topics Publishing Company, 1893). New edition published from (Huntington W. Va.: Blue Acorn Press, 2000), 21-22.

[88] New Baltimore seems to be a postwar designation. That name does not appear on the 1876 map of Heidelberg Township. This cluster of houses was located along the Abbottstown Road (Broadway) near its intersection with the Moulstown Road. In his *History of York County, Pennsylvania*, George Prowell stated that Farnsworth was near this village when the first shots were fired. Although I have not been able to locate any primary source for this information, it does seem to have merit. Since the 5th New York was in the center of Hanover at that time and two full regiments of the brigade were in front of them in the line of march, it is quite possible that the front of Farnsworth's brigade was near New Baltimore.

[89] Rebecca Scheurer stated that she "was one of the children in the throng at the Market House" as the Union troops passed through the town. She also said that when the first shots were fired, she "was pulled from between artillery horses" by her mother. See Rebecca Scheurer account.

[90] On August 24, 1863, General Meade issued General Orders No. 85, which communicated revised regulations for the organization of the ambulance corps. This order specified that each cavalry regiment was to have two ambulance wagons. It seems likely that Union cavalry regiments during the Gettysburg campaign had no more than that number. (Each artillery battery probably had one.) If all the ambulances in Farnsworth's brigade were in that same section of the column, there were likely close to a dozen ambulances in that general area when the battle broke out. Some other wagons would have been with the column also, although the great majority of Kilpatrick's train was still several miles behind at that time.

[91] According to the official reports of the 18th Pennsylvania and 5th New York, the 18th was in front of the wagons when the attack occurred. But Potter's statements indicate that at least some of the 18th was behind the wagons. There is probably some truth in both versions. The largest part of the regiment was in front of the wagons. But once the troops reached Hanover, the column

halted, and company cohesion would have been more relaxed. At that point, many of the men were dismounted and receiving food from the civilians, and likely some of the 18th were interspersed with the ambulances. It is also likely that at least one company was directly behind the wagons. When the first shots were fired and Potter's rear guard raced toward Hanover, no troops blocked their path until they reached the rear of their own regiment, which was dismounted near the southwestern edge of town. Had a line of wagons been situated along the Hanover-Littlestown Road directly in front of Potter's detail, it sees unlikely the detachment could have raced unimpeded toward Hanover. The writing of Capt. John W. Phillips (later Lieutenant Colonel) seems to confirm Potter's accounts. Phillips wrote that "...Potter was driven upon the main part of the regiment, which had reached Hanover, as stated, and had halted in the main street of the town, accepting the hospitalities of the good people of that place." Apparently, the supporting troops (possibly only one company) that Potter's detail reached were directly behind the wagons at that point. When Potter's men counterattacked, other Pennsylvanians probably raced by some of the wagons to their support. See *O. R Vol. 27, Part 1*, 1008, 1011. Also *Eighteenth Pennsylvania Regimental History*, 77, 88.

CHAPTER FOUR

[92] Potter's statement taken from an article titled "Major Potter's Recollections," *The Hanover Herald*, December 2, 1905. Hereafter referred to as Potter's Recollections, *The Hanover Herald*, December 2, 1905.

[93] *Ibid*. Potter was fairly consistent in his various postwar accounts as to the numbers of this Confederate detachment. Sometimes he stated that it numbered about sixty Southern men.

[94] Beale, *Ninth Virginia Cavalry*, 82. Beale was not the only Confederate to claim that it was the 9th Virginia that fired the first Confederate shots. Stuart's ordnance officer, John Esten Cooke, wrote that "Chambliss... at the head of the Ninth Virginia drove in their pickets...." See *John Esten Cooke*, 240.

[95] Potter's Recollections, *The Hanover Herald*, December 2, 1905. In another postwar statement (his memoirs), Potter said about 150 men joined in this counterattack. A few 5th New York men may have been a part of this number. Although the major counterattack of the 5th occurred later (after the 2nd North Carolina had reached the town), various details of ambulance guards were engaged before that time. Companies E and F of the 5th sustained several casualties of men who were on assignment to guard these vehicles and who could have been involved in the counterstrike described by Potter. However, it seems more likely these ambulance guards became casualties when the major Confederate strikes by the 13th Virginia and 2nd North Carolina reached Frederick Street (i.e., shortly after Potter's counterattack but before the larger scale 5th New York counterattack).

[96] *Ibid*. Some historians have thought that Gall's presence indicates an involvement by a significant number of the 5th New York in this early phase of the fighting. But when one examines the entire context of Potter's account, it is clear that he is referring to an initial counterstrike by members of his own regiment, not the 5th New York. Potter stated that Gall was detached from the 5th New York when the fighting started. Although Potter's statements often seem self serving, he was probably correct in this assessment. Other than any of the ambulance guards, it seems impossible that any large number of 5th New York men could have reached the area southwest of town this quickly. Also, the counterattack of the 5th occurred after the major assaults of the 13th Virginia and 2nd North Carolina had reached the center of town.

[97] *Ibid.* Some readers will note that my judgment of the performance of the 18th Pennsylvania regiment is not as harsh as many historians. I do admit to a certain home state bias, but I would suggest that very few units would have fared much better given their situation. More importantly though, an analysis of primary sources supports this contention. It is true that the 5th New York was the major force in driving the Confederates from Hanover, *after* Southern troops had reached the center of town. But accounts of officers in the 2nd North Carolina maintain that before their own regiment moved upon Hanover, an initial strike by Virginia troops had already been repulsed. Accounts of Pennsylvania officers support those of the North Carolinians in that important regard; there was at least one distinct attack and counterattack before Confederate forces reached the square. It is this initial fighting southwest of the town in which some companies of the 18th Pennsylvania were involved to a much greater degree than they have ever received credit. Although much of the regiment was routed, I believe there is more to their story than first meets the eye.

[98] *Ibid.* Potter did not state his precise location when the Confederates launched their first major assault. However, the context of his various writings seems to indicate that at least some of the Pennsylvanians were on the Westminster Road at this point. He consistently maintained that men of the 18th pushed some of the Virginians back along the "lane" from which they had originally approached, in reference to the Westminster Road. According to his account in the *Hanover Herald*, it was at about that time that his men "gave way" under the pressure of the charging Southern troops. In this particular account, Potter seemed more willing to admit that the 18th was overrun and provided more details of his own actions after the Confederates' first major attack. Some of his other writings seem self-serving and almost bitter in tone concerning this phase of the battle. Although it is speculation on my part, it might be that Potter felt that his regiment's role in the fight was ignored, while other units received the credit.

[99] *Ibid.*

[100] William H. Payne's statement from his interview with historian George Prowell. According to Prowell, this information was given to him by Payne in 1900. This account formed the basis for the article entitled "The Fight On the Pike," *The Hanover Herald*, April 25, 1908.

[101] Letter from William H. Payne to Fitzhugh Lee, May 12, 1871. Brake Collection, United States Army Military History Institute, Carlisle, Pa. Transcript on file in Gettysburg National Military Park Library. Hereafter referred to as Payne-Lee letter.

[102] Pension file of Bradley Alexander, Company E, 5th New York Cavalry, National Archives.

[103] *Eighteenth Pennsylvania Regimental History*, 90.

[104] George Hoch account.

[105] According to Cooke, after the first shots were fired Stuart "rode up rapidly." At this point, Cooke gave the general a quick assessment of the situation. See John Esten Cooke, *Wearing of the Gray*, 240.

[106] *O.R., Vol. 27, Part 2*, 695.

[107] Charles B. Thomas speech, "The Fifth New York Cavalry at Gettysburg." From *New York at Gettysburg - New York Monuments Commission for the Battlefields of Gettysburg and Chattanooga: Final Report on the Battlefield of Gettysburg, Vol. III* (Albany: J. B. Lyon Company, Printers, 1902), 1128. Hereafter referred to as Charles B. Thomas speech, *New York at Gettysburg*.

[108] Rebecca Scheurer account. Rebecca also mentioned that soldiers mistook the market house for a covered bridge.

[109] Letter of S. J. Mason to *The Hanover Herald*. Portions of the letter were included in the article entitled "Patriotic Hanover," *Hanover Evening Herald*, April 12, 1899. Hereafter referred to as S. J. Mason account.

[110] Retrospect of Rev. Ambrose Schmidt. Anthony, *The Battle of Hanover*, 136.

[111] Charles B. Thomas speech, *New York at Gettysburg, Vol. III*, 1128. The 5th New York was involved in more than one distinct charge, as stated by Major Hammond in his official report. (After the initial counterattack stalled, a second was undertaken, which forced the Confederates back to the high ground south of Hanover.) But I have not been able to locate any primary sources that indicate the specific companies involved in each. Evidence does seem to indicate that Company A was in the forefront. Thomas Burke of that company captured a Confederate flag, while Selden Wales was killed while bravely rallying his men out front. See letter dated September 4, 1863, from F. M. Sawyer, Adjutant, 5th New York Cavalry, to Mrs. Bowen, sister of Selden Wales. B. Conrad Bush, compiler, *Articles from Wyoming County Newspapers and Letters from Soldiers of the 5th New York Cavalry* (West Falls, N. Y.: Bush Research, 2000), 103-104. Hereafter referred to as Bush, *Articles from Wyoming County Newspapers and Letters from Soldiers of the 5th New York Cavalry*.

[112] Article entitled "Attack on the Union Cavalry Troops by General Stewart" [sic], *The Hanover Spectator*, December 4, 1863, reprinted from original article of July 3, 1863.

[113] George Hoch account.

[114] Reverend Zieber account.

[115] Lt. S. A. Clark, Company F, 1st Vermont Cavalry, *The National Tribune*, February 23, 1888. Hereafter referred to as S. A. Clark account.

[116] Horace Ide account.

[117] *Ibid.*

[118] *Ibid.* See also report of Lt. Col. Addison Preston, 1st Vermont, *O.R Vol. 27, Part 1*, 1012.

[119] Lt. J. P. Allum, Company B, 1st West Virginia Cavalry, *The National Tribune*, September 29, 1887. Hereafter referred to as J. P. Allum account.

[120] Payne-Lee letter. It is not certain whether the 5th New York or Companies D and M of the 1st Vermont struck first. Possibly their counterattacks were almost simultaneous.

[121] Ide account.

[122] Payne-Lee letter.

[123] According to Horace Ide (1st Vermont), "the balance of the brigade passed just through the town and formed a line facing the enemy." J. P. Allum (1st West Virginia) stated "Just after we got through the town we met the Rebels in line of battle." The accounts of 1st West Virginia men are not specific as to how they actually reached that position. They might have moved through open ground south of Abbottstown Street, then crossed York Street to reach Baltimore Street. Or, after reversing direction, they might have taken Abbottstown Street back to the square before turning south onto Baltimore Street. Above statements are from the Horace Ide and J. P. Allum accounts.

[124] In a letter dated July 22, 1863, Corp. John W. Jackson, Company F, 5th New York, wrote, "Captain Eldridge [sic] of the 4th regular Artillery soon got his four pieces in position." See R. L. Murray. *Letters from Gettysburg: New York Soldiers' Correspondences from the Battlefield* (Wolcott, NY: Benedum Books, 2005), 143. Hereafter referred to as *Letters from Gettysburg*.

[125] One puzzle encountered during my research concerns the origin of the name Bunker Hill. Even after much research and discussions with local historians, I have not been able to determine how this term came into existence. The title appears to have originated after the war. Although several newspaper articles were written on the battle in the late 1800s, the earliest reference I

have been able to find is in a *Hanover Herald* article from July 5, 1902. Because Bunker Hill has become a commonly used designation among students of the battle, I have continued to use that term.

[126] *O.R. Vol. 27, Part 1*, 1008.

[127] Horace Ide account.

[128] Letter dated September 4, 1863 from F. M. Sawyer, Adjutant, 5th New York Cavalry to Mrs. Bowen, sister of Selden Wales. See Bush, Articles from Wyoming County Newspapers and Letters from Soldiers of the 5th New York Cavalry, 103-104.

[129] George Hoch account.

[130] *O.R. Vol. 27, Part 1*, 1008.

[131] The precise location of this incident is difficult to determine. Clearly, it occurred southwest of the borough. Major Hammond's report indicates that the flag was captured as the Confederates retreated, after they had already been pushed out of town. Burke's reference to the "battery" suggests the capture may have occurred along the axis of the Westminster Road. In *The Hanover Spectator* of July 3, 1863, the incident was described as "a desperate conflict in a lane a short distance from town." In the Forney family history written in 1893, Lucy Forney Bittinger wrote, "A Union soldier captured a Confederate flag in the field beyond the house," in reference to the Karle Forney home, but no source was cited for that particular piece of information. The book *Encounter at Hanover* (page 47) has some specific information on the incident. But this section is one of the few parts of the book's battle narrative that is not taken from Prowell's *History of York County, Pennsylvania*. Unfortunately, no primary source is cited for these few sentences, and they seem to be completely at odds with the account given by Burke. For Burke's statement, see W. F. Beyer and O. F. Keydel, editors, *Deeds of Valor, From Records in the Archives of the United States Government: How American Heroes Won The Medal of Honor*, two volumes (Detroit: The Perrian-Keydel Company, *1907), Vol. I*, 218. For Forney history, see Lucy Forney Bittinger, *The Forney Family, 1690-1893* (Pittsburgh, Pa.: Shaw Brothers Press, 1893.), 47. For Hammond's report see *O.R Vol. 27, Part 1*, 1009.

[132] Beale, *Ninth Virginia Cavalry*, 83.

[133] *Ibid*.

[134] *Gillespie*, 149. Gillespie wrote that Kilpatrick was near the village of Hendrix when he began his ride back to the battle. Hendrix, however, was nowhere near the Abbottstown-Hanover Road and, in fact, was very close to the Carlisle Road. When Gillespie wrote his memoirs several years later, it is possible that he consulted a local map and mistakenly looked at the wrong road. Unfortunately, his error has been passed along by several writers.

It has generally been assumed that after his meeting at the Jacob Wirt house, Kilpatrick left Hanover at the front of the 1st Michigan. It is also possible that he remained in Hanover for a time and then rode to catch up with those troops. But if so, he must have moved quickly after leaving Hanover. Primary sources place him ahead of at least some of Custer's men when the first shots were fired. According to Gillespie, when the ammunition chest exploded in Pennington's battery (see Chapter 8), Kilpatrick was ahead of the battery and looking back in the direction of Hanover toward the approaching artillerymen. Gillespie believed the general's wild ride lasted about twenty minutes. Interestingly, this estimate fits the accounts of other Union men concerning the length of the fighting in the town. Several 5th New York soldiers claimed that the Confederates were pushed out of town in about twenty minutes. Meanwhile, accounts from men in the 1st Vermont and 1st West Virginia indicate that Kilpatrick reached Hanover just after the most intense fighting had ended and their regiments had rushed back through the town toward the battle.

[135] Most people who have more than a passing interest in the battle "know" that Kilpatrick's horse died from exhaustion after the general raced back to Hanover. The story has been repeated so often and accepted so unquestioningly, that raising doubts about its accuracy can cause surprise among students of the war. The story may be true, but an examination of its lineage reveals an example of how doubtful some of our historical beliefs can be. In the *History of York County, Pa.* (1886, edited by John Gibson), historian George Prowell described Kilpatrick's ride and stated that his horse "never afterward recovered from the fearful strain." It is apparently this cryptic statement that seems to have set in motion the whole interpretation. I have not been able to find any source on this incident, primary or secondary, written before 1886.

In the early 1900s, Prowell wrote his own *History of York County, Pennsylvania*, in which he stated that Kilpatrick's horse died "a few hours later" (after the general's arrival in town). By 1945, the story was reinforced with the publication of *The Battle of Hanover*. Much of this work's core battle narrative is actually a word-for-word reprint of Prowell's *History of York County, Pennsylvania.* However, other parts of the book contain different references to this same incident. Local historians Robert Spangler and Lida Bowman Meckley wrote sections entitled "Reminiscences" that reinforced the legend. Although Meckley does not mention Kilpatrick's horse specifically, her account of a horse buried near the Broadway/Ridge Avenue intersection has sometimes been construed as referring to that particular animal. Also, when *The Battle of Hanover* was produced, approximately 500 copies contained a supplement that included eight maps by historian Robert E. Russell. According to Russell, Kilpatrick actually "destroyed" the worn out animal; in one of his maps, Russell even marks a spot where the horse was supposedly put down.

By following these accounts, one can see the story grew somewhat in the first half of the twentieth century. The earliest reference by Prowell in 1886 seems to indicate the horse died, while his 1907 work was more specific, stating that the horse died "a few hours later." Prowell may very well have been correct. But with the publication of *The Battle of Hanover* in 1945, contradictory accounts of the event were printed in the same book! According to Spangler, the horse "fell over dead," apparently immediately upon arrival in the town, while Russell asserted that Kilpatrick actually shot the animal. Interestingly, each of these quotes cited above are secondary at best. Spangler, Meckley, and Russell were born after the war and did not cite any primary source of information for their statements.

Along with *The Battle of Hanover*, the other book that firmly "fixed" the story in the public perception was *Encounter At Hanover* (1962). According to page 48 of *Encounter At Hanover*, "The horse that carried the gallant rider to the town of Hanover died a few hours later." Like much of the book's core battle narrative, that sentence is a word-for-word reprint from the 1907 *History of York County, Pennsylvania*. Since almost every latter-day writer has taken the story from either *The Battle of Hanover* or *Encounter at Hanover*, it can be stated that what we "know" about this incident can be traced directly to George Prowell.

Kilpatrick gave some speeches in the postwar years, including a few in Hanover. I have not found the actual text of any of these speeches. Although Hanover newspapers covered the lectures, their articles mention nothing of the fate of the horse. Also, Samuel Gillespie, one of the members of Kilpatrick's escort, gave an account of Kilpatrick's ride but did not mention the horse dying. Although an absence of evidence is not proof, it seems odd that if Kilpatrick's horse had died, a man who was present at the time would not have mentioned such a noteworthy event.

[136] Several primary sources including the Charles Blinn diary and the William Payne-Fitz Lee letter indicate that Kilpatrick's headquarters was in the Central Hotel. (See also the Payne statement cited in the following footnote.) In the 1886 Gibson *History of York County, Pa.*, George Prowell wrote that Kilpatrick "located his headquarters in room 24, of the Central Hotel, while Gen. Custer was in the house of the late Jacob Wirt ..., and Gen. Farnsworth in the house of William Wirt ..." Note: at the time of this writing, an unused movie theater sits at the site of the Jacob Wirt house on Frederick Street. The William Wirt ...structure is located at the southeast corner of the square.

[137] Statements of Lt. Col. William H. Payne from an interview with George Prowell. From article entitled "Who Was the Prisoner - An Incident of the Battle of Hanover, June 30, 1863 – A Rebel Colonel In a Tan Vat," *Evening Herald*, June 29, 1901. Occasionally, confusion has resulted concerning the location of Payne's capture, and with good reason. At that time, two separate tanneries were in operation along the western edge of town, and both had owners named Forney. One of these businesses, known as the Hanover Tannery, was located at the intersection of Chestnut and High Street and was owned and operated by George Forney. (Previously, this establishment had been run under a co-partnership of George and his brother David, but in 1859, their partnership had been dissolved.) The other tannery was situated on Frederick Street and was owned by Jacob Forney but was operated at that time by Henry Winebrenner. (This is not the same Jacob Forney who owned the land on which Confederate artillery was positioned.) In a deed dated November, 27, 1875, Forney conveyed ownership of the land and buildings to Henry Winebrenner and his son David. The deed also granted to the Winebrenners the rights to water from the Forney Spring but gave some specific restrictions on its use. It was at this "Winebrenner Tannery" where Payne was taken prisoner. At least three different Union soldiers gave narratives concerning the capture of this Southern officer. Considering that a large number of troops would have been in the vicinity at the time, it is not surprising that more than one Union man would have claimed credit for capturing the lieutenant colonel. The version closest to the truth is likely that of 1st Vermont cavalryman Horace Ide. Although Ide did not personally capture Payne, he gave a description of the events that seems more realistic than most accounts. It also most closely matches the explanation given by Payne himself, which was reported in the *Evening Herald* article cited above. Payne's account by itself is an interesting study. The article provides a classic case of how a historical account must be examined in light of the narrator's viewpoint and the viewpoints of the intended audience. The article was based on an interview given by Payne to local historian George Prowell. The Southern officer knew that his description would be read by a Hanover area audience with Northern sympathies. The article is filled with humor and complimentary remarks regarding the Union soldiers, especially Kilpatrick. In contrast to this account is a letter written by Payne in 1871 to a fellow Confederate, Fitzhugh Lee. This letter, while describing the same events, has a completely different tone in some particulars, especially its much harsher assessment of Kilpatrick.

[138] *Ibid.* Concerning wounds of a more serious nature, Maj. Amos White (5th New York) and Maj. Joseph Gillette (13th Virginia) were also wounded.

[139] Dickinson was apparently the highest ranking officer killed that day. His demise is mentioned in the *Supplemental Reports*, Vol. 69, 863, and in his own compiled service record. The *Official Records, Vol. 27, Part 2*, 713-714, cite an officer killed from the 1st Virginia, rather than the 10th. This notation is possibly a transcription error. According to a few postwar Hanover newspaper articles, an officer named Davis from the 1st Virginia Cavalry was killed that day. A

[139] few men named Davis did serve with this regiment. But according to the compiled service records of the 1st Virginia, none were killed on June 30.

[140] Foster died in prison at Fort Delaware during August of that year. See compiled service record of James B. W. Foster, National Archives.

[141] Blackford, *War Years with Jeb Stuart*, 226.

[142] McClellan, *I Rode with Jeb Stuart*, 328.

[143] Samuel Forney statement. From article entitled "Some Interesting Facts – Hitherto Unpublished Incidents Of Hanover Related By Capt. Graham of North Carolina," *Record Herald*, December 3, 1904. In a separate article, which was based on an interview with Forney, it was claimed that *twenty-two* dead horses were found between the Forney and Keller residences. See the article entitled "Forney Boys Experience" from *The Hanover Herald*, July 8, 1905. Samuel's boyhood home was located near the intersection of Frederick Street and (modern day) Forney Avenue. But the farm house referred to in these accounts is more likely the one which he himself owned by the time the articles were written. That house was near the intersection of the Westminster and Hanover-Littlestown Roads, on a parcel of land that had been previously owned by his father (Karle) during the war, then later deeded to Samuel. The context of these articles suggest that these dead horses were found along the Westminster Road between the postwar Samuel Forney farm and the Keller farm, not along the Hanover-Littlestown Road.

[144] *Eighteenth Pennsylvania Regimental History*, 91.

[145] J. P. Allum account.

[146] Reverend Zieber account.

[147] Samuel Althoff narrative. From article entitled "Forney Boys Experience," *The Hanover Herald*, July 8, 1905. A few variations exist of the spelling of this priest's name. Catani is the spelling found in the book by John T. Reily (Publisher), *History and Directory of the Boroughs of Adams County* (Gettysburg, Pa.: J. E. Wible Printer, 1880), 135, 136.

[148] Samuel Forney narrative. From article entitled "Forney Boys Experience," *The Hanover Herald*, July 8, 1905. Because Reddick was from North Carolina, he was assumed to have been a member of the 2nd North Carolina. However, no man by that name is listed in service records of that regiment. Samuel's compiled service record reveals that he was actually in the 13th Virginia. When the war started, Samuel was attending the University of Virginia and joined the Virginia regiment with some of his classmates. For more on Samuel Reddick, see Roger Harell, *The 2nd North Carolina Cavalry* (Jefferson, N. C. and London: McFarland and Company, Inc., 2004), 420, note 170.

[149] S. J. Mason account.

[150] *O. R., Vol. 27, Part 1,* 992.

[151] W. W. Blackford, *War Years with Jeb Stuart*, 226.

[152] Robert J. Driver, *10th Virginia Cavalry* (Lynchburg, Va.: H. E. Howard, Inc., 1992), 39.

CHAPTER FIVE

[153] Evidence suggests that the Confederates did not hold the town for long. For the approximate time elapsed between the first shots and the Confederate retreat, see Boudrye, *Records of the Fifth New York Cavalry*, 65, also *New York at Gettysburg*, 1128, and the account of D. H. Robbins in *The National Tribune*, May 20, 1915.

[154] Article by William A. Graham entitled "From Brandy Station To The Heights of Gettysburg," *The News and Observer*, Raleigh, North Carolina, Feb. 7, 1904. Hereafter referred to as Graham, (Raleigh newspaper article).
[155] *Eighteenth Pennsylvania Regimental History*, 39.
[156] *O.R. Vol. 27, Part 1*, 1009.
[157] Primary sources are not specific concerning the position of the 1st West Virginia. Evidence seems to indicate that these men were positioned south of the square throughout the battle. At least a few of their casualties were cared for in homes on Baltimore Street. Samuel Althoff witnessed a line of Union men along Baltimore Street, but he did not state their regiment.
[158] The official report of the 1st Vermont was not specific as to the location of Bennett's battalion but stated that it was "an important position on the left of town." Horace Ide's account makes reference to a road, likely the York Road. The York Road would have been critical to guard to prevent a possible flanking movement. See *O.R. Vol. 27, Part 1*, 1012.
[159] *O.R. Vol. 27, Part 1*, 1009.
[160] S. A. Clark account. Also see *O.R. Vol., 27, Part 1*, 1012.
[161] Report of Maj. George Newcombe, 7th Michigan Cavalry, *Supplemental Records, Part 1 (Reports), Vol. 5*, 267.
[162] The official report of the 7th Michigan seems to indicate that this regiment and Pennington's Battery were initially positioned between Abbottstown and York Streets. Once the Union batteries were both situated on Bunker Hill, it is likely that Elder was east of the Carlisle Road, while Pennington was to the west. In the address of Charles B. Thomas, "The Fifth New York Cavalry At Gettysburg," he stated that Elder's Battery was "posted on a hill to the north and east of the town, on the right flank of the enemy..." See *New York at Gettysburg*, 1128. Also see Pennington's report in *Supplemental Records, Part I (Reports), Volume 5*, 284.
[163] Testimony of John Bair, Adam Klunk, John Klunk, John A. Miller, and Davis Garber, November 9, 1868, and June 17 and 26, 1882. Federal Damage Claim of John Bair, National Archives Record Group 92, Vol. 2, Book 214, Claim 1541. Census and tax records suggest that this particular John Bair lived in the Hanover Borough but owned several parcels of land outside of the Borough limits.
[164] Testimony of Alfred Michael, John Jones, John Wolf, Conrad Tate, and Lewis Tate, July 21, 1864; Oct. 23, 1868; and June 15, 1882. Federal Damage Claim of A. N. Michael, National Archives Record Group 92, Vol. 2, Book 214, Claim 1547. According to the Hanover Business Directory listed on the 1860 York County map, Michael's business was located on Carlisle Street. The map itself shows a structure owned by him near the location of the modern Public Library site.

CHAPTER SIX

[165] Josiah Thoman account, included in the article entitled "The Fight On The Pike," *The Hanover Herald*, April 18, 1908. This mill was located on the east side of the Abbottstown Road, just north of the intersection with the road now known as Maple Grove Road.
[166] *O.R. Vol. 27, Part 2*, 692. Stuart's statement is backed up by a dispatch sent by Gen. George Sykes on June 30, from Union Mills, Maryland. Sykes' Fifth Union Corps was then bivouacked on the same land where Stuart had encamped the previous evening. Sykes forwarded information to army headquarters that Stuart's force was "said to be about 5,000, with six guns." Further confirmation is given in a dispatch sent by a Union signal officer, Capt. J. H. Spencer. Spencer

forwarded information to his superior officer in Washington that Stuart's force had been sighted near Dranesville, Maryland, with six pieces of artillery. See *O.R. Vol. 27, Part 3*, 424, 397.

167 By mid-1863, Maj. Robert F. Beckham commanded the one battalion of horse artillery in the Army of Northern Virginia. Beckham's command was composed of six batteries. Whether Beckham personally accompanied Stuart's movement is not certain. Although Stuart did not state the specific batteries that were present, by process of elimination alone, it can be stated that these six guns were all/part of the batteries of Breathed and McGregor. The other batteries in Beckham's horse artillery can be accounted for and ruled out. Also, McGregor and Breathed were the only batteries that cited casualties for this particular stage of the campaign. See *O.R. Vol. 27, Part 2*, 714. There is also direct evidence of their presence with Stuart's movement from letters and statements by the following individuals in *The Bachelder Papers*: Lt. Philip Johnston, 1224-1225; Col. William Morgan, 1276; Capt. William Brown, 1331; Capt. William McGregor, 1417. (Note concerning the last source cited: In the original letter, McGregor makes reference to his battery being at Hanover. When the letter was transcribed, this sentence was omitted.) Also see sources cited in Robert J. Trout, *Galloping Thunder: The Story of the Stuart Horse Artillery Battalion* (Mechanicsburg, Pa.: Stackpole Books, 2002), 705, footnote 4. Hereafter referred to as Trout, *Galloping Thunder*.

168 Robert H. Moore II, *The 1st and 2nd Stuart Horse Artillery* (Lynchburg, Va.: H. E. Howard, Inc., 1985), 1, 3-4, 8, 16, 164, 174. Hereafter referred to as Moore, *The 1st and 2nd Stuart Horse Artillery*. See also Trout, *Galloping Thunder*, 24-27.

169 *The Bachelder Papers*, 1331.

170 In his original letter, McGregor wrote the sentence that specifically referred to the Blakely rifle being at Hanover and Carlisle. Unfortunately, that sentence was omitted in the recent printings of *The Bachelder Papers*. See letter of W. M. McGregor to Maj. H. B. McClellan, June 9, 1886. Original transcript in the John Bachelder papers, New Hampshire Historical Society. Copy in McGregor's battery file, Gettysburg Licensed Battlefield Guide Library Collection.

171 Francis Wigfall letter, July 7, 1863. William Morgan, an officer in the 1st Virginia Cavalry, confirms Wigfall's statement. According to Morgan, "…three brigades of our corps, with Breathed's battery of artillery…" were selected for the expedition. Wigfall's letter is reprinted in *Moore, The 1st and 2nd Stuart Horse Artillery*, 67. Original in the Wigfall papers, Library of Congress. Morgan's letter from *The Bachelder Papers*, 1276.

172 Statement of Henry Sell to George Prowell. Sell's account formed the basis of an article written by Prowell entitled "Reminiscences of the Civil War Times." See *The Hanover Herald*, October 14, 1905. Hereafter referred to as Henry Sell account.

173 Although the evidence is not definitive, I believe that McGregor's guns were the first artillery to arrive and fired the first cannon shots. One Confederate who wrote extensively in the postwar years was Henry H. Matthews (of Breathed's battery). Although Matthews stated specifically that Breathed moved with Chambliss's Brigade on the march to Hanover, his writings seem to indicate that Breathed did not open fire until after the Confederates had retreated from the town. But according to several accounts of civilians and Union soldiers, shells were fired even before the Confederate charges reached the center of Hanover. Unless Matthews made a serious omission concerning his own battery's participation in the battle, the evidence suggests that those first artillery shots were fired by McGregor. Matthews' account is from article entitled "Pelham-Breathed Battery: The raid into Maryland, Hanover, and Carlisle, Pa., up to and including Gettysburg," *Saint Mary's Beacon*, (Saint Mary's County, Maryland) April 20, 1905. Hereafter referred to as Matthews' account, *Saint Mary's Beacon*.

[174] Henry Matthews' writing (cited in previous note) suggests that Breathed's entire battery traveled with Chambliss's Brigade. But other primary accounts indicate that one section of Breathed's guns was in the rear of Stuart's forces. One is the account of William Zimmerman. In the summer of 1863, Zimmerman left his home in Baltimore to join the Confederate army. His brother, George, had enlisted in Breathed's battery earlier in the war. In an amazing twist of fate, Zimmerman reached Westminster in the early morning of June 30 and began to follow the Confederate column. On his way to Hanover, he struck up a conversation with a soldier in Breathed's battery. The soldier told William that two of Breathed's guns were in the front of the column and two were in the rear. The soldier further stated that George was with one of the front guns, but that he would help William find his brother when the battery reunited. Interestingly, William also wrote that he did not reunite with his brother until the Confederates reached Carlisle, which suggests that during the battle at Hanover, Breathed's battery never fought as a complete unit. For the full reminiscences of William Zimmerman, see internet site: http://www.rootseb.com/~leeg/zimmerman/zimme.htm]

[175] Matthews' account, *Saint Mary's Beacon*.

[176] Several other houses were struck by artillery fire. According to various secondary sources, these include the homes of David Althoff, Philip Britcher, Philip Breighner, and Conrad Moul.

[177] Edmund J. Raus, Jr., *Generation on the March: The Union Army at Gettysburg* (Lynchburg, Va.: H. E. Howard, Inc., 1987), 156, 160.

[178] James A. Morgan, III, *Always Ready, Always Willing: A History of Battery M, Second United States Artillery from Its Organization Through the Civil War* (Gaithersburg, MD: Olde Soldier Books, Inc., no date), 1, 46-50.

[179] Pennington's report. See *Supplemental Records, Part I (Reports), Vol. 5*, 284.

[180] The Confederate casualty figures list ten men of Breathed's battery and five of McGregor's battery missing for this portion of the campaign. Compiled service records for the ten in Breathed's unit verify their capture at Westminster. At least one from McGregor's, J. B. Lampkin, was also captured there. See *O.R. Vol. 27, Part 2*, 714. Also compiled service records of those batteries, National Archives Record Group 109, M324.

CHAPTER SEVEN

[181] According to B. J. Haden of the 1st Virginia, his regiment was at the front of Lee's column during the march. His account also indicates it was the 1st Virginia that made the initial strike against the 6th Michigan. Haden's memoirs were originally written as newspaper articles, then reprinted in book form. See B.J. Haden, *Reminiscences of J.E.B. Stuart's Cavalry* (Progress Publishing Co: Charlottesville, VA, 1912). His original memoirs have been more recently edited by Timothy A. Parrish and reprinted in the publication of the Fluvanna County, Virginia Historical Society called *The Bulletin*, Issue Number 55, Spring, 1993. See section entitled "Raiding in Pennsylvania,". Haden's accounts hereafter referred to as B. J. Haden memoirs, *The Bulletin*.

[182] The Snyders and Waltmans resided a few miles north of Littlestown.

[183] Solomon and William Snyder testimony, October 23, 1868. State Damage Claim of Henry Waltman.

[184] There is substantial evidence from civilian accounts that once Lee's men reached Pennsylvania, they used the Pine Grove Road and other nearby farm lanes. Lee himself confirmed the final part of the march in a dispatch written to Stuart (See note 11, this chapter). But the initial portion of their screening movement from Union Mills is more difficult to interpret. An examination of

contemporary maps shows that the Maryland roads connecting with the Baltimore-Littlestown Pike eventually led back to the path of the main column on the Hanover Road. If Lee's men turned off the Pike before they reached Pennsylvania, they would have been forced to merge with the main column for a short time before moving off to the left once again, or they would have needed to use isolated farm lanes leading north from Humbert Schoolhouse Road to reach Pennsylvania. To rejoin the main column would have created a major impediment on an already congested road and would have been an unacceptably inefficient way of protecting the column. No Southern accounts have surfaced of any of Fitz Lee's troops causing any delays or even of coming close to the main column before reaching Hanover. The evidence from various civilian damage claims suggests two likely possibilities for this stage of the movement. Lee might have turned onto Humbert Schoolhouse Road, then used a farm lane to move into Pennsylvania before striking Clouser or Bart's Church Road. Or, he might have reached Pennsylvania on the Baltimore-Littlestown Pike before turning east on Mathias Road. By taking either of these routes, the screening force would have remained on a parallel course with the main body for a longer time and would have had smaller patrols roam other side roads to act as an early warning system. (It seems that Lee's main body did not use Bollinger Road, although it might have been traveled by a few small detachments. No damage claims were filed by any civilians from Bollinger Road, and that course would have taken Lee away from the Pine Grove Road, which saw considerable Confederate activity, as recorded by multiple witnesses.)

[185] George Bair, David Bair, and Henry Sterner testimony, Nov. 7, 1868. State Damage Claim of George Bair. Bair's testimony did not reveal the numbers of Confederates that he saw nor the direction from which they rode onto his property. Whether Bair saw Lee's full brigade or a small detail is in question, and is what helps to make this segment of Lee's ride the most difficult to establish of all the Confederate movements that day. Clouser Road is shown on the 1872 Union Township map but not the 1858 Adams County map. However, it is apparent that some secondary roads and farm lanes not marked on the 1858 map were in existence at that time. Many home sites are indicated on the 1858 map, with no roads shown near them. Line Road, although not noted on the 1858 map and only partially shown on the 1872 Union Township map, was clearly in existence at the time as, oddly enough, it shows on the northern edge of Carroll County, Maryland maps of both 1862 and 1872. Confederates might have reached Bair's farm via a farm lane now called Biemiller Road. Although not shown on the 1862 and 1872 Carroll County maps, a few farm sites are shown along its path, indicating the likelihood of the lane's existence at the time.

[186] Jeremiah Sellers testimony, November 7, 1868. State Damage Claim of Philip Fase. Sellers himself had a horse seized. See Emanuel Wildasin testimony, October 23, 1868. State Damage Claim of Jeremiah Sellers.

[187] Adam Brumgard and Philip Fase testimony, November 7, 1868. State Damage Claim of Philip Fase.

[188] It cannot be stated definitively that it was troops of Lee's Brigade that took Fase's horse. However, the proximity of this location to Lee's route strongly suggests that scenario. On the other hand, the Fase claim reveals a great benefit of these civilian accounts. Fase specified that he was near the Jacob Lohr mansion on the public road to Hanover when the incident took place. (This house is shown on the 1858 Adams County map.) Locals often had specific knowledge of the landmarks, roads, and geographical features of the areas in which they lived and worked. Several encounters with citizens created a paper trail of testimony on Confederate movements, although it seems that not nearly as many horses were taken by Lee's Brigade as by detachments

from Stuart's main column. With their primary job as a protective force, possibly Lee's men did not actively roam in search of horses but only gathered those close by or within sight. Also, this section of Union Township was possibly the poorest area of south central Pennsylvania at that time; the agricultural census figures indicate fewer horses in this area than other nearby townships.

[189] William Wisensale and Michael H. Kitzmiller testimony, October 23, 1868. State Damage Claim of William Wisensale.

[190] John Panebaker and Peter Panebaker testimony, October 22, 1868. State Damage Claim of Peter Panebaker.

[191] This message was found on a captured Confederate courier and later delivered to Kilpatrick. A reporter who was traveling with the general used the dispatch in a newspaper article. See article entitled "Our Cavalry Battles," *The Philadelphia Inquirer*, July 4, 1863.

CHAPTER EIGHT

[192] Report of Col. Russell Alger, 5th Michigan Cavalry, *Michigan in the War*, 578.

[193] From the speech of James Harvey Kidd, (captain, Company E, 6th Michigan Cavalry) given June 12, 1889, at the dedication of the Michigan monuments at Gettysburg. See *Michigan at Gettysburg*, 138. Also see report of Col. George Gray in *Michigan in the War*, 580.

[194] *Michigan at Gettysburg*, 138-139. Although Kidd believed that it was close to noon when the citizen reported the information, it is possible the incident actually occurred before that time. If the first indication of a Confederate presence was the word of this citizen, it suggests two possibilities: Either the regiment began its movement to Hanover before the first cannon shots were fired, or the 6th was in an area where the men could not hear the sounds of artillery fire. Kidd believed the troops spotted by the citizen were Lee's Brigade. It is possible that he assumed this in hindsight since it was Lee's men who attacked the 6th Michigan. But an individual five miles out on the road toward Hanover might have seen members of Chambliss's Brigade southwest of Hanover.

[195] *Ibid.*, 139.

[196] Report of Col. George Gray, *Michigan in the War*, 580.

[197] *Michigan at Gettysburg*, 139 (J. H. Kidd speech). Also see letter of Daniel Stewart (Company B, 6th Michigan) to Margaret Murray, July 24, 1863. From Regional Historical Collections, Western Michigan University, Kalamazoo, Michigan. Copy in 6th Michigan Cavalry file, Gettysburg National Military Park Library collection. Hereafter referred to as Daniel Stewart letter.

[198] B. J. Haden memoirs, *The Bulletin*, 21. Lee's leading regiments may have already established contact with Stuart's main body by the time the 6th Michigan came into their view, but Haden's account seems to indicate otherwise. Apparently, the 1st Virginia was still moving into position on Chambliss's left when Lee spotted the 6th Michigan.

[199] Report of Col. George Gray, *Michigan in the War*, 580.

[200] Daniel Stewart letter.

[201] Letter from John B. Kay to his brother, July 10, 1863. Bentley Historical Library Collection, University of Michigan. Copy in 6th Michigan Cavalry file, Gettysburg National Military Park Library. Hereafter referred to as John Kay letter to brother.

[202] Letter from John B. Kay to his parents, July 6, 1863. Bentley Historical Library Collection, University of Michigan. Transcript in 6th Michigan Cavalry file, Gettysburg National Military Park Library.

[203] Compiled service records of Adam, George, and Isaac Bare, 1st Virginia Cavalry, National Archives. See also Robert J. Driver, Jr., *1st Virginia Cavalry* (Lynchburg, Va.: H. E. Howard, Inc., 1991), 149. Hereafter referred to as Driver, *1st Va. History*.

[204] B. J. Haden memoirs, *The Bulletin*, 21.

[205] Report of Col. George Gray, *Michigan in the War*, 580.

[206] Accounts of Company B state that they used wooded area(s) in this covering action. The site cannot be stated with certainty, and it is possible that not all the charges took place in the same location. A few likely areas, particularly for the initial counterattack, are the patches of woods west of Narrow Drive, between the Schwartz Schoolhouse and Gitt's Mill. The largest of these forested areas seems to be the most probable site. Aerial photographs taken in the 1950s and terrain maps from the 1930s indicate that the dimensions of these woods did not change substantially between then and the present day. Property lines indicated in deeds of the late 1800s and early 1900s suggest the woods were also present then. They were likely present in those areas in 1863, although the exception may be the small patch of woods on the Schoolhouse grounds. One piece of evidence that should be considered is from the *Record of Service of Michigan Volunteers,* in which ten men are listed as missing from Company B on June 30 but none from Company F. This item strongly suggests that the two companies became separated from each other during their holding action or were initially positioned in different areas. It is possible that it was men of Company F who raced southward toward Littlestown and made contact with the 5th Michigan along the Hanover-Littlestown Road. Weber was killed on July 14, 1863, at Falling Waters, Maryland. He had just been promoted two days before that. In separate reports, Custer and Kilpatrick refer to him being a major at the time of his death. However, in the casualty lists for the Gettysburg Campaign, he is cited as being a captain when he fell. Possibly the promotion had not yet been formally recognized, and Weber was still officially a captain while acting on assignment as major.

[207] Letter of Daniel H. Powers to his parents, July 19, 1863. Hereafter referred to as Daniel H. Powers letter. This letter is the property of David Van Dyke of Nappanee, Indiana. I am much in debt to the Van Dyke family for allowing me to use this account and to Mark Stowe of Grand Rapids, Michigan, for contacting the Van Dyke family concerning its use. Mark had utilized this account in his excellent history, *Company B - 6th Michigan Cavalry*, a self published work that I came across in the U. S. Army Military History Institute in Carlisle, Pennsylvania.

[208] *Ibid.*

[209] *Ibid*

[210] Allen D. Pease letter to his mother, July 18, 1863. This letter is the property of Eloise Haven of Grand Rapids, Michigan. I am much in debt to her for allowing me to use this account and once again to Mark Stowe, who put me in contact with Eloise.

[211] Daniel H. Powers letter.

[212] Daniel Stewart letter.

[213] *Ibid.* It is not known where Weber's detachment was hidden after they fought the covering action. Considering that some of the 5th Michigan were south of the Hanover-Littlestown Road as they made their way to Hanover, but did not make contact with (at least some of) Weber's men, it seems that his detachment was trapped well south of that road.

[214] John Shaeffer and Edward Shaeffer testimony, October 22, 1868. State Damage Claim of John Shaeffer. Shaeffer's land straddled the York/Adams County line; his farmhouse was located several hundred yards east of Narrow Drive, on land that is now part of the South Hills Golf Course. Since Shaeffer claimed all the damage done to his land was by Confederate skirmishers, while Jeremiah Gitt and Samuel Schwartz files claims for damages done by Union soldiers, their statements provide an interesting basis for speculation as to how close Michigan troops were able to get to the Confederate rear and where the skirmish lines were located in this area.

[215] Testimony of Samuel Schwartz, Jeremiah Gitt, and John Zimmerman, October 22, 1868 and February 5, 1883. Samuel Schwartz Federal Damage Claim, National Archives Record Group 92, Book 2, Vol. 214, Claim 733.

[216] Some of this fighting was within one and a half miles from McSherrystown; at least one man from the 3rd Virginia was said to have been wounded at McSherrystown. See Thomas P. Nanzig, *3rd Virginia Cavalry* (Lynchburg, Va.: H. E. Howard, Inc., 1989), 38. Also see Driver, *1st Va. History*, which lists an index of the men who served in the regiment. The compiled service records of Confederates in the National Archives, however, often contain minimal information. There are several men of the 1st Virginia who became casualties but whose compiled service records do not reveal this information.

[217] Diary of Lt. Col. William R. Carter. From Lt. Col. William R. Carter, C.S.A., *Sabres, Saddles, and Spurs*, edited by Col. Walbrook D. Swank, U.S.A.F., retired. (Shippensburg, Pa.: Burd Street Press publication, printed by Beidel Printing House, Inc., 1998), 73. Hereafter referred to as Carter Diary, *Sabres, Saddles, and Spurs*.

[218] Letter of Maj. Luther Trowbridge, 5th Michigan Cavalry, United States Army Military History Institute, Carlisle, Pennsylvania. Copy in 5th Michigan Cavalry file, Gettysburg National Military Park Library. Hereafter referred to as Luther Trowbridge letter.

[219] John Allen Bigelow account, from article entitled "Draw Saber, Charge!" *The National Tribune*, May 27, 1886. Hereafter referred to as John Allen Bigelow account. This soldier had enlisted early in the war with the 1st Michigan Cavalry under the name John Allen Bigelow. He later reenlisted with the 5th Michigan Cavalry under the name John Allen. In one account, it was claimed that Bigelow was the first member of the 5th Michigan to use a saber in battle, in what was referred to as the "incident" at Littlestown, Pennsylvania. See *Biographical Record: Biographical Sketches of Leading Citizens of Oakland County, Michigan* (Chicago: Biographical Publishing Company, 1903), 611.

[220] John Allen Bigelow account.

[221] *Ibid.*

[222] Luther Trowbridge letter. Trowbridge and John Allen Bigelow were two soldiers who made a clear differentiation between the 5th Michigan fighting and that done closer to Hanover.

[223] Testimony of David Boyer, Emanuel Wildasin, and William Wisensale, January 2 and 4, 1869, and June 3, 1884. Federal Damage Claim of Emanuel Wildasin, National Archives Record Group 92, Vol. 2, Book 214, Claim 738. Neither Boyer, Wildasin, or Wisensale mentioned the specific regiment(s) they saw, but several facts indicate it must have been the 5th Michigan. All of Boyer's land was in Union Township and well west of the land where the 6th Michigan fought. Any involvement here by Farnsworth's brigade can also be ruled out, as none of his regiments deployed any large numbers of their men before reaching Hanover. Also, if Farnsworth's troops had skirmished here, Union generals would have been aware of the Confederate presence well before reaching Hanover. The troop estimate of one thousand Union soldiers given by Boyer and Wisensale was likely way off. But clearly this unit was not a small

[224] *Ibid.* Wildasin was a tenant on land owned by David Boyer. Boyer owned parcels of land on both sides of the Hanover-Littlestown Road. Testimony given in this damage claim and in the Pennsylvania State Claim of Jeremiah Sellers seem to indicate that Wildasin resided south of the Hanover-Littlestown Road. The most important piece of evidence concerning the position of Union troops in Wildasin's rye field is the statement he makes that the field was "on a hill." The only hills on Boyer's land were south of the Hanover-Littlestown Road. Since this particular section of the road has changed course, the hill he mentioned might be the one the modern road traverses near the Sheppard Road intersection, or the higher ground south of there.

[225] Testimony of Levi Maus, August 12, 1881, and Lewis Carbaugh, August 13, 1881. Federal Damage Claim of Levi Maus, National Archives Record Group 92, Vol. 2, Book 214, Claim 782.

[226] John Allen Bigelow's account states that a full battalion of the 5th Michigan was ordered to dismount and move to the right of the road. Maj. Crawley Dake wrote that "four squadrons dismounted as skirmishers, the remaining four squadrons in reserve, mounted..." It is likely he used the word "squadrons" in reference to companies. But both the Bigelow and Dake accounts indicate that approximately half of the force was eventually dismounted. This deployment was critical to control the road itself. In 1863, the road crossed the Conewago Creek a few hundred yards north of the present day bridge on Route 194. The section of the road for several hundred yards on the west side of the crossing (in Union Township) was on much lower ground than the modern road trace. That portion of the Union route was dominated by the higher ground on which some of the present road now travels. In Bigelow's account, he stated that the 5th Michigan initially pushed the Confederates back about two miles before they found greater numbers of the enemy. This distance corresponds to the account of B. J. Haden of the 1st Virginia. Haden wrote that when his unit made first contact with Union troops (the 6th Michigan east of Schwartz Schoolhouse), they charged for at least two miles before they ran into other Union reinforcements (likely the 5th Michigan.) See Bigelow account. Also report of Maj. Crawley P. Dake, 5th Michigan in *Supplemental Records, Part 1(Reports), Vol. 5*, 261.

[227] Report of Colonel Alger, *Michigan in the War*, 578.

[228] Confusion has occurred regarding the location and ownership of this site. Hanover merchant Josiah Gitt lived in the borough but owned some parcels of land outside of the town, including a farm along the Westminster Road. Josiah, however, did not own the mill where the skirmishing took place. This mill was owned by Jeremiah Gitt and situated in Adams County. (See Union Township Tax Assessment Books in the Adams County Historical Society, along with contemporary maps and the damage claim of Jeremiah Gitt cited below.) The mill is no longer in existence but was located a few hundred yards south of the intersection of Narrow Drive and Lovers Drive. The site is now private property.

[229] Although most of Fitz Lee's men rode right past the mill, the horses may have been taken by soldiers from a different brigade. The mill site is less than one mile from the main column route on the Westminster Road. Testimony of Jeremiah Gitt and Casper Krepps, October 21 and 23, 1868 and August 21 and 23, 1882. Federal Damage Claim of Jeremiah Gitt, National Archives Record Group 92, Vol. 2, Book 214, Claim 743.

[230] *Ibid.* Casper Krepps, who worked on the Gitt property, also witnessed this skirmishing. Krepps verified that it was Michigan troops who occupied the mill.

[231] The initial contact between Lee's Brigade and the 6th Michigan was made within one mile of this mill. Before long, elements of these units began to spill over onto surrounding farms and

woodlands. But that charge took place north of the Gitt site and pushed at least most of the 6th away from the area. *If* 6th Michigan troops occupied the mill, it was likely part of Weber's detachment, either during their holding action or later when they were cut off behind the Confederate lines. However, of the 6th Michigan accounts which I have been able to locate, none mention a mill. Possibly the mill was utilized by 5th Michigan men. By the time the 5th reached this vicinity, much of Lee's Brigade had moved or been pushed back further eastward. The John Allen Bigelow account and the testimony of David Boyer, William Wisensale, and Emanuel Wildasin indicate that large numbers of Union troops were deployed south of the Hanover-Littlestown Road. The road known today as Sheppard Road (now partly a private road) was a wartime road and ran along David Boyer's land directly to Gitt's Mill. See John Allen Bigelow account and the Emanuel Wildasin Federal Damage Claim.

[232] This account is from the article entitled "An Incident of the Battle of Hanover," *Hanover Evening Herald*, June 30, 1903. The blacksmith shop was on land that, in postwar years, became the Martin Arnold farm. Several questions remain about this incident. In the *History of York County, Pa.* (edited by John Gibson, 1886), George Prowell mentions that the first casualty on June 30 was a Confederate officer who was killed about three and a half miles southwest of Hanover. Apparently, Prowell was referring to the same incident mentioned by William Gitt. In the *Official Records*, the only Confederate cavalry officer cited as being killed during this portion of Stuart's expedition is an unnamed officer of the 1st Virginia. A few postwar newspaper articles claim that a Capt. J. A. Davis of that regiment was killed at Hanover. However, in my own research, I have not been able to find any indication to confirm either the citation in the *Official Records* or the newspaper accounts, even after an exhaustive search through all the compiled service records of the 1st Virginia. One Confederate officer who was killed that day was Capt. James Dickinson of the 10th Virginia. Possibly the citation in the *Official Records* is a transcription error, with 1st mistaken for 10th. According to his compiled service record, Dickinson was killed in the town itself, but this seems unlikely. Another possibility is that the soldier killed near Gitt's Mill was not an officer. It has been assumed that this casualty occurred before any major action occurred at Hanover. That is possible, but the soldier might have been shot as part of the fighting which took place when the 6th and 5th Michigan moved toward Hanover. Gitt's land bordered that of both Samuel Schwartz and John Shaeffer and was just south of the area where first contact was made between the 6th Michigan and Lee's Brigade (See appendix J). Shooting also likely broke out on all of their properties again when the 5th Michigan moved through the area. Another puzzle is how William, as a young boy, would know the location of the grave, while his father, who owned land not far from the site, would know nothing of the burial spot. William's statement is also intriguing in another regard. Since it seems unlikely that he had witnessed the incident, how he knew where the man had been shot is not certain. Local oral history holds that the soldier was shot close to the mill. The cavalryman was possibly killed near the mill then buried near the blacksmith shop several hundred yards away. Or, the soldier may have been wounded near the mill and made it back to Conewago Hill, then died near his burial spot. According to some secondary sources, the shot which killed this Confederate was fired from the Drescher farm. Although at least one Drescher is listed in the 1863 tax rolls of Union Township, Adams County, this reference would seem to be the name of a postwar owner, and likely refers to a farm on the Fairview Road in York County, several yards east of the Adams County line. The *Official Records* citation listed above is from *Vol. 27, Part 2*, 713. Information on the oral history of the Gitt family was given to the author by Peter and Sharon Sheppard, the current owners of the land where Gitt's Mill stood. Sharon is a descendent

of the Gitt family. Information on the Drescher farm location was given by Wanda L. Gerber, a former employee of the Sheppard family and a descendent of the Drescher family.

[233] There are indications that Union movement(s) caused Stuart concern for the safety of the captured wagons. Some writers believe that the dismounted advance by the 6th Michigan (after they reached Hanover) was the cause of this fear. However, with large numbers of Southern troops holding a very dominant position south of town, it is difficult to see how the movement of a dismounted skirmish line in the Confederate front could have caused Stuart much alarm. Graham's statement about an advance in the "rear" suggests a much more plausible scenario. The threat occurred from the fighting closer to the Schwartz Schoolhouse and/or Gitt's Mill area. The movement of Michigan troops in those areas took place much closer to the captured wagons. This statement taken from the Graham (Raleigh newspaper article) cited in chapter 7, note 2.

[234] From speech by Eugene Dumont Dimmick at a reunion in Hanover for veterans of the 5th New York Cavalry, June 30, 1913. (Held in conjunction with the fiftieth anniversary celebration of the Battle of Gettysburg.) Extracts of this speech were printed in an article entitled "The Official Report," *Hanover Record Herald*, July 2, 1913. In this newspaper article, the account is mistakenly labeled as being that of Col. Nathaniel Richmond, but the events described indicate that it is clearly Dimmick's account.

[235] It is extremely difficult to determine the units involved and the exact timing of these types of encounters. While most of the shooting away from the primary battle area took place after the outbreak of the "main" engagement at Hanover, certainly the possibility exists of a few encounters by scouts before that time. When a civilian reported skirmishing, the location can often be ascertained without knowing what units were involved. When soldiers' accounts relate the fighting, the units are known, but the locations are often doubtful. The statements of Wentz and Esale raise the possibility that at least some of Wentz's farm changed hands during this fighting. Neither farmer mentioned the numbers of troops they had seen. Some of David Boyer's property was less than a mile to the north, so it is possible that the fighting on the Wentz land involved 5th Michigan men on the fringes of that skirmish. However, small details from almost any unit could have been involved here. Federal Damage Claim of Jesse Wentz, National Archives Record Group 92, Vol. 2, Book 214, Claim 739.

CHAPTER NINE

[236] *O.R. Vol. 27, Part 1*, 825.

[237] *Ibid.*, 835-836.

[238] Account of Charles Morse, 2nd Massachusetts Infantry. From a speech given May 10, 1878, and later included in the regimental history. 2nd Massachusetts Infantry file, Gettysburg Licensed Battlefield Guide Library.

[239] Letter of Allen Rice, Company C, 6th Michigan Cavalry. Rice did not specify the name of the town, but the context of his account provides overwhelming evidence that the location of this skirmishing was just outside of Littlestown. Transcript on file in 6th Michigan Cavalry file, Gettysburg National Military Park Library.

[240] Account of Joseph Lumbard, Company G, 147 Pennsylvania Infantry. Lumbard's diaries were published in the *Snyder County* (Pennsylvania) *Tribune* in weekly installments from 1876 through 1878. The only complete copy known to exist is in the John P. Nicholson collection (manuscript archives) of the Henry Huntington Library in California. The copy used here was

taken from a microfilm roll in the collection of Susan Boardman, Licensed Battlefield Guide, Gettysburg, Pennsylvania. Hereafter referred to as Joseph Lumbard account.

[241] Henry C. Morhous, *Reminiscences of the 123 N. Y. S. V.* (Greenwich, N. Y.: Peoples Journal Book and Job Office, 1879), 46-47. Copy on file in Gettysburg Licensed Battlefield Guide Library. Hereafter referred to as Morhous, *123 N. Y. History*.

[242] Letter of Lt. Robert Cruikshank, Company H, 123 New York Infantry. From manuscript at Bancroft Public Library, Salem, New York. Copy on file in Gettysburg Licensed Battlefield Guide Library.

[243] *Ibid.*

[244] Joseph Lumbard account.

[245] Morhous, *123 N. Y. History*.

CHAPTER TEN

[246] *O.R. Vol. 27, Part 2*, 695. See also H. B. McClellan, *I Rode With Jeb Stuart*, 328. As of this time, I have not been able to find any primary sources to indicate the order in which Hampton's regiments marched toward Hanover.

[247] Confederate unit locations are impossible to pinpoint with certainty. On the Confederate left flank, it is likely that Lee's Brigade did not remain in the area for long before it began to escort the wagon train. It is even possible the brigade never actually held a battle-line position on the Confederate left. The fact that the 6th Michigan was able to advance a line of dismounted skirmishers south of the Hanover-Littlestown Road without instigating a major counterattack suggests that Lee had left the area by that time. The name Beck Mill is a shortened version of Beckers Mill Road. In the 1860 census, Levi Becker is listed as the master miller of this business. The mill was located along the south branch of the Conewago Creek.

[248] Once Hampton's men were in position, the Confederate line stretched approximately two miles with close to five thousand men. It is likely that significant concentrations of troops were positioned near the artillery units and the roads. In particular, the Westminster Road, Baltimore Pike, and York Road had large numbers of troops positioned within several hundred yards. Southern artillery also fired from on or near these same roads.

[249] Confederate artillery was positioned on the high ground near Mount Olivet Cemetery. That fact can be firmly established by examining various civilian accounts. But over the years, the story has even occasionally included Southern artillery firing from among the tombstones. This supposition is highly unlikely, to say the least. At that time, the cemetery was only a fraction of its present size, and this area south of town offered great fields of fire from many acres of high ground on both sides of the Baltimore Pike. It seems incredible that any officer would set up artillery pieces in an area with obstructions (like tombstones) to the loading and firing process, when a move of several yards in any direction would have easily avoided this difficulty. (The cemetery was laid out into lots in 1859, under the direction of the Mount Olivet Cemetery Association. By the end of 1862, eighty-four burials had taken place and probably close to one hundred by the time of the battle. See article entitled "Mt. Olivet Cemetery," *The Hanover Citizen*, April 29, 1875.)

[250] Statement of Reverend Zieber from article entitled "The Battle of Hanover," *The Hanover Herald*, July 15, 1905. Also see official report of General Kilpatrick in *O.R. Vol. 27, Part I*, 992.

[251] Testimony of Samuel Mumma, Jacob Bart, January 10, 1868, State Damage Claim of Jacob Bart.

[252] It is possible that artillery was in this area for only a very short time. Reverend Stock's statement was from a speech given in 1900 at a celebration of the thirty-seventh anniversary of the battle. See speech of Rev. Dr. Charles Stock, reported in *The Hanover Herald*, July 7, 1900.

[253] According to various contemporary maps, more than one Forry lived in this general area. Two neighbors, John M. Hershey and John S. Hershey, were able to witness the events on Forry's land from their property. This fact seems to indicate that this particular Forry residence was the one located near the modern intersection of Center Street and Spring Avenue. See evidence of John M. Hershey, John S. Hershey, and George T. Forry, January 15, 1869, State Damage Claim of George T. Forry.

[254] *O. R. Vol. 27, Part 1*, 1012.

[255] Horace Ide account.

[256] In Major Hammond's report, he stated that that when the effort was made to flank the Confederate position, he was also instructed to "order an advance of the skirmishers on the right...". He did not specify who these skirmishers were. They possibly were part of his own regiment who were to support a mounted advance by other 5th New York troops. Or, he may have been referring to the battalion of the 1st Vermont under Major Bennett. It is possible the attack on the Confederate flank was to be a coordinated assault, with the 5th New York mounted and the Vermont battalion dismounted. Also, Hammond did not state the location of the Confederate battery which he was to capture. If there was Confederate artillery near the York Road at that time (see notes 251 and 252 above), certainly the advance of the 5th New York would be toward that position. It seems unlikely that a single regiment would have been ordered to assault the high ground near the Mount Olivet Cemetery, which was an easily defensible position with the troops Hampton had in that area. Hammond's statement taken from *O.R Vol. 27, Part 1*, 1009.

[257] John Kay letter to his brother. Kay wrote that the 5th Michigan was deployed on foot but did not state where they were positioned in relation to the 6th Michigan. The writings of Capt. James H. Kidd confirm that the 5th Michigan had reached Hanover before the 6th made its dismounted advance. See J. H. Kidd, *Personal Recollections of a Cavalryman: With Custer's Michigan Cavalry Brigade in The Civil War* (Ionia, Mich.: Sentinel Printing, 1908. Reprinted by Grand Rapids, Mich.: The Black Letter Press, 1969), 128. Hereafter referred to as J. H. Kidd, *Recollections*.

[258] Henry Sell account. When historian George Prowell adapted Sell's interview for his newspaper article, he wrote that two regiments were dismounted at this point. According to Prowell, one [the 6th Michigan] "stood in line of battle from St. Matthew's Church northward to the Ketterer Wagon Works." (This wagon works was located where the present day railroad tracks cross Elm Avenue.) At this point, they "wheeled around and moved forward to the Littlestown Road..." Prowell then wrote that "Another regiment [likely the 5th Michigan] of dismounted men took position in a line extending from St. Matthew's Church toward McSherrystown."

[259] J. H. Kidd, *Recollections*, 128.

[260] *Ibid*, 129.

[261] Henry Sell account.

[262] Forney's first name did have an "e." This spelling was taken from his mother's maiden name, "Karle." This spelling can also be found on various primary sources such as deeds, census records, and damage claims. Testimony of Karle Forney, June 13, 1882. Karle Forney Federal Damage Claim, National Archives Record Group 92, Vol. 2, Book 214, Claim 1525.

[263] John Esten Cooke, *Wearing of the Gray*, 242. The most critical factor on this part of the field was likely the decision to have Lee's Brigade escort the wagon train. As long as Lee's men remained

on the Confederate left, Union control of the Hanover-Littlestown Road was questionable at best, even with the majority of the 6th Michigan south of the road. Once Lee withdrew his brigade, the road was once again secure for Union communications.

[264] Kilpatrick referred to "forming a junction with the main army, from which we had been separated for hours." *O.R. Vol. 27, Part 1*, 992.

CHAPTER ELEVEN

[265] The Westminster, Beck Mill, Black Rock, York, and Blooming Grove Roads, along with the Baltimore Pike were all in existence at this time and shown on the 1860 York County map.

[266] *O.R. Vol. 27, Part 2*, 696. Also H. B. McClellan, *I Rode with Jeb Stuart*, 329.

[267] Testimony of Sarah, Joseph, and Martin Arnold, October 23, 1868. State Damage Claim of Joseph Arnold. This farm is on Fairview Road but well east of the Westminster Road. It is unlikely that any large body of Confederate troops would have moved through Arnold's fields before the retreat from Hanover, with the possible exception of the wagon train itself. When Hampton's Brigade moved toward the battle, it likely continued on the Westminster Road well north of the Arnold farm before turning eastward. Even if Hampton had turned onto Fairview Road to reach the Confederate right flank, he could have then taken Beck Mill Road directly to the Mount Olivet area without moving through Arnold's crop fields.

[268] Ephraim Nace account. Fairview Road is the only road in this township that leads directly from the area where the wagons were parked to the Baltimore Pike.

[269] *Ibid*. Larry Wallace, Hanover resident and Gettysburg Licensed Battlefield Guide, has researched and uncovered a wealth of information on the engagement at Hanover. Larry stated that when he was teaching at West Manheim Elementary in the mid 1960s, the Centre School House still stood at the northeast corner of the Baltimore Pike and Fuhrman Mill Road.

[270] Ephraim Nace account. Nace did not mention this farmer's name but stated that the incident occurred "soon after we passed Centre school-house." Local historians, including Clark Wentz in his *History of West Manheim Township*, have associated this skirmish with the Anthony Brockley farm. Brockley filed a state damage claim in which his son John gave testimony. John stated that he was attempting to remove his father's horses to safety, when eight or nine Confederates seized the horses. He further said, they took him prisoner "until Union forces under General Kilpatrick came up" when they told him to leave as there was going to be a fight. Most likely these witnesses and historians were all referring to the same incident. Most secondary writings of this event state that the boy who revealed the location of the soldiers was twelve years old. Of Brockley's sons, the closest to this age would have been Joseph, who was listed as eleven in the census taken in August of 1860. Yet the actions of the boy during this incident seem to suggest one even younger, who did not understand the danger of the situation. Possibly the most likely candidate is Urias, who was listed as four years old in the 1860 census. The identity of the Union soldiers involved is also a mystery, but one possibility is that they were from the 1st West Virginia. When the 1st West Virginia took position south of the center of town, likely a detachment was sent south of Hanover to scout along the Baltimore Pike. These troops might have been cut off when Hampton took position closer to the town. The official report of the 1st West Virginia cited eighteen men captured, a curious number considering that the unit only listed seven men killed or wounded. Another possibility is involvement of 5th New York troops. In a letter dated July 22, 1863, Cpl. John W. Jackson, Company F, 5th New York, wrote that after the Confederates withdrew from Hanover, "...A detachment from our regiment under Major

Hammond followed them four miles, skirmishing with their rear guard." See *Letters from Gettysburg*, 143. Testimony in the Anthony Brockley State damage claim was given November 26, 1868. The local historical work cited above is Clark B. Wentz, *History of West Manheim Township* (1969 unpublished manuscript, copy in vertical file of the Pennsylvania Room, Guthrie (Hanover) Public Library.), 40. Hereafter referred to as Wentz, *History of West Manheim Township*.

[271] William W. Allbright and Henry Rudisill testimony, November 2, 1868. State Damage Claim of William W. Allbright.

[272] Simon Barnhart testimony, October 23, 1868. State Damage Claim of Simon Barnhart.

[273] The "Wildasin" lane is now a private drive. It is possible that upon reaching the end of Fuhrman Mill Road, some of the column turned right onto Black Rock Road and a short distance later turned left onto Dubs Church Road. This option, however, is less direct and seems less likely.

[274] John Wildasin, Levi Christman, and Christian Millheim testimony, October 23, 1868, and January 27, 1869, from state damage claims of the same individuals. The specific reference to the time of day is from the testimony of Christian Millheim in the claim file of Levi Christman. In a separate incident less than a mile away, Catherine Wildasin had three horses seized. Andrew, Catherine's son, said the time was "soon after noon" when the animals were taken. Millheim's testimony, in particular, seems to indicate that these Confederates were detached from Fitz Lee's column, which passed by after they left Hanover. The horse gathering details sent out earlier from Chambliss would not likely have ridden that far east while their brigade was still at Hanover. Captain Graham's detail made it back to where the wagon train was parked before Fitz Lee's Brigade even completed its screening movement from Union Mills.

[275] West Manheim historian Clark Wentz gave the 265 estimate for the three townships. George Prowell believed that 385 horses were seized from Codorus Township alone. Although estimates by twentieth century historians vary, it is clear that a staggering number of animals were seized. According to the Pennsylvania Auditor General's Department, 686 York County residents filed claims for damages inflicted by Confederate forces. The great majority of these losses were horses taken by Stuart's forces, with many claimants losing more than one animal. See *Notes On The Index To Damage Claim Applications By Auditor General's Department Officials, 1894*. From microfilm, Adams County Historical Society, and Pennsylvania State Archives. For estimates of local historians, see Wentz, *History of West Manheim Township*, 40. Also George R. Prowell, *History of York County, Pennsylvania* (Chicago: J. H. Beers and Company, 1907), 430.

[276] The most critical road addition in this area is the stretch of Route 216 between the Codorus State Park office and the southern tip of the lake. The original route from Blooming Grove to Glen Rock veered off the modern road path near where the park office stands today and continued across the lake area. Not far from the Dubs Church, the road intersected another road and ran northeast for a short distance. The road then turned south and continued through the area that is now the southern arm of the lake, where the original trace once again "rejoins" modern Route 216.

[277] The road known today as Dubs Church Road did not actually go directly to the Church in 1863. Before it reached the area where modern Route 216 is situated, the road made a right turn (into what is now a wooded area that is part of the State Park) and intersected another road, which led north past the church and toward the village of Smith's Station. Some of the original trace to Smith's Station is now under the lake; other parts of that route can be found in areas of Codorus State Park. It is likely that the column followed the Dubs Church Road for its entire 1863 length; turned left onto the road to Smith's Station; then several hundred yards later, turned right at the

church. The section of the road that ran northeastward from the church is also no longer in use. Traces of this road still exist, however, leading from the church toward the area of the boating marina. However, another possibility exists for the troops to reach the Dubs Mill area without going directly by the church. It is possible that after using the Dubs Church Road (or farm lanes east of there), the column continued to the village of Marburg and moved north to Dubs Mill, then east toward Jefferson. A few facts, however, suggest against this latter scenario. To reach Marburg from Dubs Church Road, the column would have had to turn southward and go completely out of its way for a time or would have had to cut across significant stretches of woods or farmland. Also, eyewitness testimony seems to indicate that the only Confederates that were between Dubs Church Road and Marburg were small detachments, while large bodies of troops were present near the Dubs Church and Dubs Mill.

[278] Testimony of William Dubs and Levi Snyder, October 23, 1868. State Damage Claim of William Dubs.

[279] Testimony of John H. Snyder, Levi Snyder, and John Wildasin, October 23, 1868. State Damage Claim of John H. Snyder. The number of wagons needed to load 692 bushels (seized from Dubs and Snyder) indicates that this was no small detachment. The testimony of Levi Snyder seems to indicate that this grain was taken at about the same time Dubs' items were taken, possibly from the same mill. Less than a half mile to the south of Dubs' Mill was the village of Marburg. The remains of this village are now under the water of the lake that bears its name. Although there were other mills closer to and south of Marburg, I have not found evidence that these mills sustained losses. The 1860 census lists three men named John Snyder in Manheim Township. Several variations on the spelling of their last names can be found. However, since Snyder is the most common spelling found in various tax records, contemporary maps, and census rolls, that is the spelling I have used.

[280] *Encounter at Hanover*, 97. Luisa was the wife of George Dubs.

[281] Traces of this original road can still be found, where it leaves the lake area to the east of the "round island."

[282] Henry Rennol Testimony, January 20, 1869. State Damage Claim of Henry Rennol.

CHAPTER TWELVE

[283] George W. Beale, letter to his mother, July 13, 1863. From: *Lieutenant Of Cavalry In Lee's Army* (Baltimore, MD: Butternut And Blue, 1994), 114. Hereafter referred to as George Beale letter.

[284] Another possibility is that some of Chambliss's men took Clover Lane directly to the Baltimore Pike. Clover Lane is shown in the 1876 *Atlas of York County* (Heidelberg Township map), although portions of its course have changed since that time. The area today known as Penn Township was part of Heidelberg Township at that time.

[285] Sarah Huggins's testimony, January 22, 1869. State Damage Claim of John M. Wildasin. This John Wildasin of Heidelberg Township is not the same man as the previously mentioned John Wildasin of West Manheim Township. This site is now also under Lake Marburg.

[286] Elizabeth Stover, Ephraim Leppo testimony, January 14, 1869. State Damage Claim of Ephraim Leppo. These statements may have been referring to a part of the larger action closer to, but still east of, the Hanover Borough. But the wording and context of the testimony seem to indicate skirmishing around Blooming Grove itself. As of this writing, I have not been able to determine whether the area was actually called Blooming Grove in 1863. A church known as the Blooming

Grove Church was founded in the 1880s and is the earliest reference I have been able to find with that particular name.

[287] It cannot be stated with certainty that this 5th New York detachment was the one involved in the shooting at Blooming Grove. The 5th was positioned east of Hanover when the Confederates withdrew, so it is likely that its patrols would have moved out the York Road then past the Blooming Grove area. For Jackson's account, see *Letters from Gettysburg*, 143.

[288] Testimony of Amanda, Barbara, and Michael Bowman; George and David Pressel; Jacob and Moses Miller; and Henry and Emanuel Rennol, January 14, 20, 21, and 22, 1869. State Damage Claims of Michael Bowman, Jacob Miller, George Pressel, and Henry Rennol. Rennol was the only one of these claimants who stated a time frame when his animals were seized. He stated that this incident happened between 3:00 and 3:30, which would seem to indicate detachments from Fitz Lee's Brigade. Concerning the other landowners, Confederate patrols from Lee or Chambliss might have reached those properties along Hoff Road after they detached from the main column route south of there. The other possibility is that details from Hampton's Brigade rode the length of Hoff Road after leaving Blooming Grove Road.

[289] Testimony of John Miller, February 20, 1869. State Damage Claim of John Miller.

[290] Several civilian accounts confirm the statement of Lt. George Beale that the wagon train was well on its way from Hanover before Chambliss and Hampton left the battlefield. According to Christian Millheim and Andrew Wildasin, large numbers of Confederates had reached Manheim Township by early afternoon. Farther eastward, William Brinkman stated that he was riding his horse near the borough of Jefferson when he was overtaken by four Rebel soldiers at about 4:00. William Carter, an officer of the 3rd Virginia, wrote that his men reached Jefferson "at 6 ½ P. M." Peter Klinedinst, who resided just north of New Salem, testified that Confederates had reached his premises by about 7:00 P. M. At the least, the vanguard of Lee's Brigade must have begun the northward turn toward New Salem before nightfall. Statements of Millheim and Wildasin taken from State Damage Claims of Levi Christman and Catherine Wildasin, respectively. Testimony of Brinkman and Klinedinst from State Damage Claims filed under their own names. Carter's statement taken from Carter Diary, *Sabres, Saddles, and Spurs*.

[291] John Esten Cooke, *Wearing of the Gray*, 243.

[292] Armand Glatfelter, *The Flowering of the Codorus Palatinate: A History of North Codorus Township, Penna.* (York, Pa.: Mehl-ad associates, by authority of North Codorus Sesquicentennial Commission), 264. Hereafter referred to as Glatfelter, *History of North Codorus Township*.

[293] Anthony, *The Battle of Hanover*, 154. Also testimony of Amos Lough, David Peters, Elias Slagle, and Charles and Henry Diehl. October 21 and 22, 1868, and January 9, 1869. State Damage Claims of David Peters, Elias Slagle, Charles Diehl, and Henry Diehl.

[294] Glatfelter, *History of North Codorus Township*, 264, 265.

[295] Garland C. Hudgins and Richard B. Kleese, editors, *Recollections of an Old Dominion Dragoon: The Civil War Experiences of Sgt. Robert S. Hudgins II - Company B, 3rd Va. Cavalry* (Orange, Va.: Publisher's Press, Inc., 1993), 81-82.

[296] Testimony of John E. Ziegler, John K. Ziegler, Israel K. Ziegler, Andrew Straughsbaugh, George Harmin, and Jacob Harmin, October 26, 1868, November 8, 1871, April 25 and May 8, 1882. Federal Damage Claim of John E. Ziegler, National Archives Record Group 92, Vol. 2, Book 214, Claim 1505.

[297] George Beale letter.

[298] John Esten Cooke, *Wearing of the Gray*, 243-244.

[299] Ephraim Nace account.

CHAPTER THIRTEEN

[300] According to Pennington's report, his battery changed position a few times that day. Pennington stated that after the enemy's guns were withdrawn, his command "moved out on the road to Littlestown. Finding that the enemy had retreated, we countermarched and encamped at Hanover." Meanwhile at about five o'clock, Companies C, H, and E of the 7th Michigan, under Major Newcombe, "were sent to occupy the town." These three companies possibly were ordered to remain in support of Battery M, even while the rest of the regiment was deployed elsewhere at the time. (The report of Col. Charles Town suggests that the 1st Michigan remained near Bunker Hill even after the artillery was moved. See *Supplemental Records, Part I (Reports), Vol. 5*, 257, 267, 284.

[301] According to Major Hammond's report, the 5th New York, "went into bivouac outside the town." Col. Nathaniel Richmond claimed his regiment encamped "near" the town. Lt. Henry Potter stated that the 18th Pennsylvania camped on the "outskirts of the town." Meanwhile, Maj. Crawley Dake stated that the 5th Michigan "bivouacked in Hanover." A few locals cited damages from Union cavalry encampments, including Andrew Rudisill and David Slagle. Rudisill's farm was located about one mile northeast of the square and to the east of Abbottstown Street and (modern) Moul Avenue. For statements of Hammond and Richmond, see *O.R. Vol. 27, Part 2*, 1005, 1009. Potter's statement taken from the Henry Potter account cited previously. Dake report from *Supplemental Reports, Part I (Reports), Vol. 5*, 261. Civilian accounts from testimony of David Slagle, George Slagle, Walter Norwich, and George Klinefelter in the Federal Damage Claim of David Slagle. Also testimony of Andrew Rudisill, Jacob Miller, Jacob Bechtel, and Peter Treiber in the Federal Damage Claim of Andrew Rudisill. From National Archives Record Group 92, Vol. 2, Book 214, Claims 1522 and 1543.

[302] When the casualties for Stuart's cavalry were submitted, the figures were listed according to the interval of the campaign in which they occurred. In this case, for the period they were "on the march from Rector's Cross-Roads [the start of Stuart's expedition] to Gettysburg, and including the battle at Hanover, Pa." The numbers indicate that during this segment of the campaign, Lee's Brigade sustained forty-seven casualties, Hampton fifteen, and Chambliss thirty-seven. (The report does list some men missing from the batteries of Breathed and McGregor, but compiled service records indicate that almost all, if not all, of these men were captured at Westminster on June 30, after Stuart's forces moved to Hanover.) This citation appears not to include the action at Hunterstown, in which Cobb's Legion alone lost at least thirty men. But in this particular report, the 2nd North Carolina, the unit that sustained the greatest Confederate losses at Hanover, is not listed. Later, in a "recapitulation" report, the 2nd North Carolina is cited as having sixty killed, wounded, or missing. (This later citation included the totals for the entire campaign, but it appears that the 2nd sustained only one casualty at Gettysburg itself.) Even taking into account a few losses of the 4th Virginia in the Westminster engagement, these numbers would seem to indicate that Stuart sustained at least 150 casualties at Hanover. This estimation also coincides with Kilpatrick's report, in which he indicates that the Confederate losses were somewhat less than his own. See *O.R Vol. 27, Part 1*, 992. Confederate casualties listed in *O.R. Vol. 27, Part 2*, 713-714, 718.

[303] Testimony of Peter, Thomas, Jerome, and Michael Noel, Jesse Haar, John Myers, and Sebastian Wise, December 20, 1865, October 24, 1868, and April 27, 1882. Federal Damage Claim of Peter Noel, National Archives Record Group 92, Vol. 2, Book 214, Claim 1529.

[304] Testimony of John A. Myers and Andrew Orendorff, October 26, 1868, and May 18, 1882. Federal Damage Claim of John A. Myers, National Archives Record Group 92, Vol. 2, Book 214, Claim 1528.

[305] Rev. Ignatius Bellwalder was the legal representative of Rev. J. B. Catani (deceased). According to his testimony, the bay horse belonged to Catani, "to whom all the property of Conewago Church and Missions belonged, as Superior of the Society of Jesus." When Catani died, Bellwalder succeeded him as Superior. The horse was taken from the stable of Samuel Brady, who occupied a farm attached to the Paradise Church. Federal Damage Claim of Rev. Ignatius Bellwalder, National Archives Record Group 92, Vol. 2, Book 214, Claim 736.

[306] *O.R., Vol. 27, Part 2*, 709.

[307] *Ibid.*

CHAPTER FOURTEEN

[308] *The Hanover Spectator*, July 10, 1863.

[309] Reverend Zieber account. Also see Samuel Althoff account. Also George Prowell, *History of York County, Pennsylvania (*Chicago: J. H. Beers and Company, 1907), 439. Hereafter referred to as Prowell, *History of York County.*

[310] Prowell, *History of York County*, 440. Also see article entitled "The Battle Of Hanover," the *Evening Herald*, June 30, 1896. Also article entitled "Forney Boys Experience," *The Hanover Herald,* July 8, 1905, concerning the burial area of Samuel Reddick.

[311] Article entitled "Hospitals Removed," *The Hanover Spectator*, July 17, 1863. According to this newspaper article, this building had been vacant for some time before the battle.

[312] Letter to the editor by Surgeon Perrin Gardner, *The Hanover Citizen*, July 9, 1863.

[313] Article entitled "Our Hanover Ladies," *The Hanover Spectator*, September 11, 1863. Original letter written by Thomas Walker to the *Lapeer* (Michigan) *Republican*, August 26, 1863, entitled "The Patriotic Ladies of Hanover, Pennsylvania." The soldier was an artilleryman in a United States battery. He was likely wounded at Gettysburg and then transported to Hanover.

[314] Article entitled "Reminiscences of the Civil War Times," *The Hanover Herald*, July 21, 1906.

[315] Article entitled "Strangers," *The Hanover Spectator*, July 17, 1863.

[316] Letter in the federal pension file of Eber Cady, National Archives.

[317] *O. R Vol. 27, Part 1*, 25-26.

EPILOGUE

[318] *O.R Vol. 27, Part 1*, 70.

APPENDIX A

[319] *O.R. Vol. 27, Part 3*, 912-913.

[320] Edwin P. Coddington, "Prelude to Gettysburg: The Confederates Plunder Pennsylvania," *Pennsylvania History* (Vol. 30, No. 2, April 1963), 125-127, 153, 154, 156, 157.

APPENDIX B

[321] Robert M. Utley, *Frontier Regulars: The United States Army and The Indian, 1866-1891* (New York: Macmillan Publishing Company, Inc. and London: Collier Macmillan Publishers, 1973), 36 (endnote 8).

APPENDIX C

[322] John Gibson, historical editor, *History of York County, Pa., Part II* (Chicago: F. A. Battey Publishing Company, 1886), 61. Hereafter referred to as Gibson, *History of York County, Pa.* See also Prowell, *History of York County*, 1059.

[323] Article entitled "The History of Pennville," *Hanover Record Herald*, Tuesday, November 11, 1919.

[324] Anthony, *The Battle of Hanover*, 152.

[325] Robert G. Carter, *Four Brothers in Blue* (Norman, Ok.: University of Oklahoma Press, 1999), 298.

APPENDIX D

[326] Parsons, *National Tribune* article, August 7, 1890.

[327] *O.R., Vol. 27, Part 1*, 1018.

[328] S. A. Clark account, *The National Tribune*, February 23, 1888. See also Charles Blinn diary.

[329] Horace Ide account.

[330] Original article entitled "Pennsylvania's Patriotism," *St. Albans Daily Messenger*, 1887, (unknown month and date), reprinted in *Encounter at Hanover*, 168.

[331] J. P. Allum account.

APPENDIX E

[332] Henry Sell account.

[333] 18th Pa. Cavalry Regimental History, 427.

[334] Spangler's map was included in *The York Gazette and Daily* of July 29, 1936.

[335] Anthony, *The Battle of Hanover*, 145.

APPENDIX F

[336] Payne-Lee letter. Copies are on file in Brake Collection, Unites States Army Heritage and Education Center, Carlisle, Pennsylvania, and Gettysburg Licensed Battlefield Guide Library Files.

[337] Busey and Martin, *Regimental Strengths and Losses at Gettysburg*, 121.

[338] *Supplemental Records, Vol. 48*, 60, 61.

[339] *O.R. Vol. 27, Part II*, 718.

[340] Payne's statement was reprinted in article entitled "Forgotten Warrior" (John Coski, ed.), *North and South* magazine, Vol. 2, Number 7 (September, 1999), 85-86.

[341] Payne-Lee letter.

[342] From article entitled "Some Interesting Facts," *Hanover Record Herald*, December 3, 1904.
[343] From article entitled "Reminiscences Of The Civil War Times," *The Hanover Herald*, October 14, 1905.
[344] Busey and Martin, *Regimental Strengths and Losses at Gettysburg,* 217.

APPENDIX G

[345] Widow's pension file of Elizabeth (Lizzie) Sweitzer Waltz. Lizzie was the widow of Levi Waltz, Company D, 19th Unites States Infantry. National Archives pension file certificate number 351,338.
[346] *Ibid.*
[347] *Ibid.*

APPENDIX H

[348] Letter to the editor by surgeon Perrin Gardner, *The Hanover Citizen*, July 9, 1863.
[349] From a historical sketch of the town of Hanover, *Hanover Citizen*, July 6, 1876.
[350] Gibson, *History of York County, Pa., Part I*, 215.
[351] Article entitled "The Battle of Hanover," *Hanover Evening Herald*, June 30, 1896.

APPENDIX I

[352] Report of Col. George Gray, *Michigan in the War*, 580.
[353] J. H. Kidd, *Recollections*, 127.
[354] Henry Sell account.
[355] Testimony of Samuel Schwartz, Jeremiah Gitt, and John Zimmerman, October 22, 1868, and February 5, 1883, Samuel Schwartz Federal Damage Claim (claim number 733). Also see testimony of Solomon Schwartz and William Shaeffer, October 23, 1868, and February 5, 1883, Solomon Schwartz Federal Damage Claim (claim number 732). Both claim files can be found in National Archives Record Group 92, Book 2, Vol. 214.
[356] Testimony of Samuel, John, Henry, and Ephraim Keller, October 24, 1868 and January 14, 1882. Federal Damage Claim of Samuel Keller, National Archives Record Group 92, Vol. 2, Book 214, Claim 1524. Among many other items, Keller even listed a coat taken worth $4 and one grain bag worth $1 yet did not refer to any fighting on his land.
[357] Testimony of John Shaeffer and Edward Shaeffer, October 22, 1868. State Damage Claim of John Shaeffer. Shaeffer's land straddled the York/Adams County line; his farmhouse was located several hundred yards east of Narrow Drive, on land that is now part of the South Hills Golf Course.

APPENDIX J

[358] *Eighteenth Pennsylvania Regimental History* (account of Henry Potter), 87.
[359] Prowell, *History of York County*, 426.
[360] *Eighteenth Pennsylvania Regimental History* (account of Thomas J. Grier), 39.

361 Testimony of Jeremiah Gitt and Casper Krepps, October 21 and 23, 1868, and August 21 and 23, 1882, Federal Damage Claim of Jeremiah Gitt, National Archives Record Group 92, Vol. 2, Book 214, Claim 743.

362 John Allen Bigelow account. Also testimony of David Boyer, Emanuel Wildasin, and William Wisensale, January 2 and 4, 1869, and June 3, 1884. Federal Damage Claim of Emanuel Wildasin, National Archives Record Group 92, Vol. 2, Book 214, Claim 738.

APPENDIX K

363 W. W. Blackford, *War Years with Jeb Stuart*, 225. Also Beale, *Ninth Virginia Cavalry*, 80.

364 Prowell, *History of York County*, 428.

365 Anthony, *The Battle of Hanover*, 146. Henry's last name is spelled Gotwald in the 1860 census and Gotwalt in the 1870 census.

366 Testimony of Samuel, John, Henry, and Ephraim Keller, October 24, 1868, and January 14, 1882. Federal Damage Claim of Samuel Keller, National Archives Record Group 92, Vol. 2, Book 214, Claim 1524. The Kellers cited damage done by Union infantry encamping on the land July 1 and Confederates taking horses on June 30. Among many other items, Keller even listed a coat taken worth $4 and one grain bag worth $1 yet did not refer to any Confederates having wagons on his land.

367 Heidelberg Township Tax Assessment Books. See also Deed Ledger Book 12-D, page 206 (York County Archives). This deed gives the boundaries of the thirty-eight-acre tract, which in 1899, still had the same dimensions as when it was owned by Graby in 1863.

368 Heidelberg Township Tax Assessment Books. See also Deed Ledger Book 13-G, page 205 (York County Archives). This deed gives the boundaries of the fifteen-acre tract, which in 1903, still had the same dimensions as when it was owned by Gotwalt in 1863. Deed Book 13-G, page 111, indicates the parcels Gotwalt bought after 1863, which were east of the Westminster Road.

369 John and Edward Shaeffer testimony, October 22, 1868. State damage claim of John Shaeffer.

370 The portion of Grandview Road that is west of the Baltimore Pike is not present on the 1860 York County map. The 1876 Heidelberg Township map shows a farm lane, which partially followed Grandview's present-day course. Cooper Road is often placed on modern maps of the Battle of Hanover but is not present on 1860 or 1876 maps. Some of the western portion of Cooper Road was the southern boundary of land owned by Jacob Forney in 1863 (and later owned by Jesse Rice). The eastern portion of the road was the southern boundary of Jacob Forry. The common border apparently was used as a right of way, and sometime after the Civil War, it eventually was developed into a road.

371 Clark B. Wentz, *History of West Manheim Township*, 39. Vertical file, Pennsylvania Room, Hanover Public Library. Wentz, like Spangler, did not cite a source for his information.

Index

1st Delaware, 24, 25
1st Michigan, 9, 30, 36, 37, 57, 59
1st Ohio, 9, 30, 124
1st South Carolina, 10, 78
1st Vermont, 9, 29, 30, 36, 37, 38, 47, 48, 49, 50, 57, 59, 79, 105, 122
1st West Virginia, 9, 29, 36, 37, 38, 47, 48, 49, 50, 54, 57, 78, 97, 105, 106

2nd North Carolina, 7, 10, 32, 33, 41, 42, 44, 47, 49, 53, 55, 56, 108, 109, 110, 125
2nd South Carolina, 10, 78

3rd Virginia, 10, 71, 125, 126

4th Virginia, 10, 25, 127

5th Michigan, 7, 9, 29, 30, 36, 59, 66, 70, 72, 73, 74, 75, 81, 105, 118, 121
5th New York, 3, 9, 27, 29, 30, 36, 38, 39, 42, 44, 45, 47, 49, 50, 55, 57, 75, 79, 90, 105, 121, 123

6th Michigan, 3, 7, 9, 30, 36, 59, 65, 66, 67, 70, 71, 72, 73, 74, 75, 76, 81, 82, 84, 95, 105, 115, 116, 117, 118, 121, 125, 127

7th Michigan, 9, 29, 30, 36, 37, 57, 59, 117

9th Virginia, 7, 10, 32, 39, 41, 42, 50, 52, 89, 118, 123, 125
10th Virginia, 10, 32, 52, 53, 55, 124
13th Virginia, 7, 10, 22, 32, 39, 41, 42, 49, 50, 55, 123, 124
18th Pennsylvania, 9, 29, 36, 38, 39, 41, 42, 44, 49, 50, 54, 56, 57, 94, 98, 105, 112, 114, 115, 117, 121

35th Virginia Battalion, 17, 18, 91
123rd New York, 77
147th Pennsylvania, 4, 77

Albright's Hall, 97
Alexander, Sgt. Bradley, 42
Alger, Col. Russell, 9, 72, 73
Allbright, William W., 85
Allum, Lt. J. P., 54, 106
Arnold, Joseph, 84
Bair, George, 63
Bair, John, 59
Bare, George, 69
Barker, Capt. Augustus, 75
Bart, Jacob, 79
Battery E, 4th U. S. Artillery, 9, 29, 36, 50, 62
Battery M, 2nd U. S. Artillery, 9, 29, 36, 60, 62
Baublitz, John, 33
Beale, Lt. George W., 89, 92, 123
Beale, Col. Richard, 10, 39, 52, 110, 118, 120, 123
Bennett, Maj. John, 57, 79
Bigelow, John Allen, 72, 118
Blackford, Capt. William W., 7, 23, 54, 55, 118, 120, 123
Blinn, Charles, 30, 37, 105
Blooming Grove, 89, 90
Boudrye, Chaplain Louis, 30, 123
Boyer, David, 72, 73
Breathed, Capt. James W., 10, 61, 62, 78, 106
Brockley farm, 78, 84
Brown, Capt. Wilmer, 61
Bunker Hill, 50, 57, 59, 62, 89, 119
Burke, Pvt. Thomas, 50, 52

Cady, Eber, 98
Candy, Col. Charles, 76
Capehart, Maj. Charles, 49, 105
Carbaugh, Lewis, 73
Carter, Robert G., 104
Carter, Lt. Col. William, 71, 124
Catani, Father John B., 54, 95
Chambliss, Col. John, Jr., 10, 20, 22, 31, 32, 33, 36, 38, 44, 50, 54, 56, 57, 59, 61, 63, 67, 78, 81, 82, 89, 90, 91, 95, 115, 116
Conewago Hill, 36, 74, 75, 84, 89, 117, 118, 120
Cooke, Capt. John Esten, 42, 91, 92, 124
Cruikshank, Lt. Robert, 77
Custer, Brig. Gen. George A., 4, 7, 9, 28, 29, 30, 36, 57, 59, 81, 89, 94, 95, 99, 125, 126

Dickinson, Capt. James, 53
Diehl, Peter, 91
Dimmick, Lt. Eugene, 75
Dodge, 2nd Lieut. Winchester, 59
Dover, 83, 91, 92, 93, 95
Dubs, William and Luisa, 85
Dubs (St. Paul's) Church, 85, 89, 90
Duttera, John, 36

Early, Maj. Gen. Jubal A., 16, 17, 30, 93, 95
Eckert's Concert Hall, 97
Elder, Lt. Samuel, 9, 29, 36, 38, 50, 57, 59, 62, 79
Elder, Squire John, 60
Esale, Benedict, 75

Farnsworth, Brig. Gen. Elon J., 9, 28, 29, 30, 36, 37, 38, 48, 50, 52, 56, 57, 89, 99, 105, 106, 126
Fase, Philip, 63, 65
Forney, Jacob, 8, 107, 108, 119
Forney, Karle, 54, 55, 81, 97
Forney, Samuel and John, 36
Forry, George, 79
Forry, Jacob, 8, 107
Foster, James B. W., 53
Frank, Peter, 18

Gall, Lt. Alexander, 39
Gardner, Dr. Perrin, 97, 98, 112
Geary, Maj. Gen. John W., 76, 77
Geiselman, Daniel, 17
Gillespie, Samuel, 52, 124
Gitt, Abdiel, 17
Gitt, Jeremiah, 73, 117, 118
Gitt, Josiah, 33, 73
Gitt, William, 73, 117
Gitt's Mill, 3, 65, 67, 73, 74, 75, 117, 118
Graham, Capt. William A., 7, 33, 56, 74, 109, 110
Gray, Col. George, 9, 66, 67, 69, 70, 81, 115, 116
Greenswalt, Sgt. Jacob, 47

Haden, Sgt. Benjamin J., 69, 70
Ham, Joe, 92
Hammond, Maj. John, 9, 45, 50, 52, 56, 57, 79, 90
Hampton, Brig. Gen. Wade, 10, 20, 22, 31, 32, 35, 56, 57, 63, 75, 78, 79, 82, 89, 90, 91, 92, 95, 96, 99
Hanover Commons, 14, 15, 17, 45, 47, 49
Hartman's mill, 60
Hoch, Sgt. George, 36, 42, 45, 47
Hoff, Henry, 91
Hoffacker, John, 47, 112, 113, 114, 115
Hollinger, David, 60
Hudgins, Sgt. Robert, II, 92, 125
Huggins, Sarah, 89

Ide, Horace K., 37, 48, 49, 50, 79, 105, 122

Jackman, Lt. S. J., 54

Jackson, Cpl. John, 90
Jefferson, 82, 84, 85, 88, 91, 92

Kay, John, 69
Keller, Samuel, 8, 35, 54, 61, 106, 107, 116, 119, 120
Keyes, Sgt. William, 70
Kidd, Capt. James H., 66, 67, 81, 115, 116, 125
Kilpatrick, Brig. Gen. Hugh Judson, 6, 7, 9, 11, 26, 27, 28, 29, 30, 31, 36, 37, 39, 41, 52, 53, 55, 56, 57, 59, 62, 65, 66, 71, 76, 78, 79, 82, 83, 92, 94, 95, 97, 99, 100, 104, 110, 122, 127
Kitzmiller, Michael, 65
Kitzmiller's Mill, 36
Knap's battery, 77
Koiner, C. H., 70
Krepps, Casper, 117, 118

Lake Marburg, 85, 89
Lee, Brig. Gen. Fitzhugh, 7, 10, 20, 21, 22, 24, 25, 31, 32, 56, 59, 63, 65, 66, 67, 71, 72, 73, 75, 78, 82, 88, 89, 90, 91, 95, 108, 109, 110, 115, 116, 118
Lee, Gen. Robert E., 6, 12, 13, 15, 16, 17, 19, 20, 22, 23, 24, 26, 27, 93, 94, 95, 99, 100, 101, 102, 103, 122, 123, 125
Leese, Adam, 33
Leib, Joseph, 18
Leppo, Ephraim, 90
Leppo, Jacob, 35, 36
Litchfield, Lt. Col. Allyne, 57
Lohr, Jacob, 65
Lovell, Lt. D. G., 95
Lumbard, Joseph, 77

Macomber, Dexter, 30
Marion Hall, 97
Market House, 42, 47
Market Square, 97
Mason, S. J., 44, 55
Maus, Levi, 73
McClausland, Billy, 70
McClellan, Maj. Henry B., 7, 54, 78, 126
McElwain, Lt. John, 30
McGregor, Capt. William M., 10, 61, 62
Meade, Maj. Gen. George G., 27, 66, 79, 94, 100, 126
Michael, Alfred, 59
Miller, John, 90
Millheim, George, 85
Moran, James, 60, 62

Mount Olivet Cemetery, 78, 114
Mount Pleasant, 67, 69, 115, 116
Mumma, Samuel, 79
Myers, John, 95

Nace, Ephraim, 35, 84, 93
New Salem, 91, 92
Noel, Peter, 94, 95

Panebaker, John and Peter, 65
Payne, Lt. Col. William H. F., 10, 41, 42, 49, 53, 56, 108, 109, 110, 122
Peale, Sgt. Isaac, 54
Pease, Allen, 3, 71, 125
Pennington, Lt. Alexander C. M., Jr., 9, 29, 36, 50, 57, 59, 60, 62
Phillips, " Bang", 92
Phillips Legion, 10, 78
Pigeon Hills, 13, 31, 44
Pleasant Hill Hotel, 97, 98
Pleasonton, Maj. Gen. Alfred, 26, 27, 28
Potter, 2nd Lt. Henry, 39, 41, 105, 117
Powers, Lt. Daniel, 3, 70, 71

Reddick, Samuel, 55
Reformed Church, 37, 97
Rennol, Henry, 88
Rice, Allen, 76, 77
Rice, Jesse, 106, 107, 108
Rickey, Corporal, 50, 52
Rife, John, 17
Rudisill, Henry, 85

Scheurer, Rebecca, 44
Schmidt, Ambrose, 44
Schwartz, Samuel, 66, 71, 116, 118
Schwartz, Solomon, 66, 116
Schwartz Schoolhouse, 3, 4, 66, 67, 71, 72, 73, 74, 75, 78, 115, 116, 118
Sell, Henry, 36, 61, 81, 106, 115, 116
Seven Valleys, 91, 92
Shaeffer, John, 67, 71, 116, 118, 119
Shields, Lt. Thomas, 42, 56
Shriver, Andrew, 31
Shriver, Herbert, 32
Shriver, William, 31
Shriver Mill, 32

Smith, Doctor, 97
Smith, Capt. J. B., 42
Snyder, William and Solomon, 63
Snyder, John, 85
Spangler, George, 37
Spencer Rifles, 29, 72, 73, 81, 82
St. Matthew's Church, 15, 81, 97
Stahl, William, 17
Stewart, Daniel, 71
Stock, Rev. Dr. Charles, 79
Stover, Elizabeth, 90
Stuart, Maj. Gen. J.E.B., 4, 6, 7, 10, 11, 16, 19, 20, 22, 23, 24, 25, 31, 32, 33, 34, 35, 36, 39, 41, 42, 44, 45, 52, 53, 54, 56, 59, 61, 62, 63, 65, 66, 67, 71, 74, 75, 76, 78, 81, 82, 83, 84, 85, 89, 91, 92, 93, 94, 95, 97, 99, 100, 103, 109, 110, 118, 119, 120, 122, 123, 126, 127

Tate, Conrad and Lewis, 59
Thoman, Josiah, 60
Trone, Daniel, 17
Trowbridge, Maj. Luther, 72
Twelfth Corps, 66, 76, 77, 82, 94, 114

Union Mills, 4, 25, 31, 63, 66, 104

Vollum, Lt. Col. Edward, 98

Wales, Sgt. Selden, 50
Waltman, John and Nathaniel, 63
Weber, Capt. Peter, 70, 71, 73
Weber, Major, 118
Wentz, Jesse, 75
White, Lt. Col. Elijah, 17, 18, 91
Wildasin, Emmanuel, 73, 118
Wildasin, John, 85, 89
Winebrenner, D. E., 107
Winebrenner, Henry, 62
Winebrenner Tannery, 53, 109
Wirt, Calvin, 37
Wirt, Jacob, 37
Wisensale, William, 65, 72

Zeigler's Church, 91
Zieber, Rev. William K., 37, 47, 54, 78, 97
Ziegler, John, 92
Ziegler's Mill, 92

About the Author

The author is a native of McSherrystown, Pennsylvania and is a descendent of four ancestors who fought in the Civil War. Three fought with the following infantry regiments: the 48th PA., the 91st PA., and the 202nd PA. Another one, Michael Krepps, was a member of Battery G, 1st PA Light Artillery. Michael was wounded at the Battle of Gaines Mill and taken prisoner. He returned to his unit and fought at Gettysburg on East Cemetery Hill on July 2, 1863 when the unit came under attack.

The author grew up and still resides on land that was once controlled by Confederate troops during the Battle of Hanover. He has been a Licensed Battlefield Guide at Gettysburg since 1995. He has previously written on Confederate cavalry movements for Blue and Gray magazine, and also a short history of his ancestor in Battery G, 1st Pennsylvania Light Artillery.

****

Other Books by Colecraft:

Civil War Artillery at Gettysburg **by Philip M. Cole**

Command and Communication Frictions in the Gettysburg Campaign **by Philip M. Cole**

Human Interest Stories of the Gettysburg Campaign **by Scott L. Mingus, Sr.**

Human Interest Stories of the Gettysburg Campaign –Vol.2 **by Scott L. Mingus, Sr.**

Human Interest Stories from Antietam **by Scott L. Mingus, Sr.**

A Concise Guide to the Artillery at Gettysburg **by Gregory A. Coco**

Remarkable Stories of the Lincoln Assassination **by Michael Kanazawich**

For Ordering Information:

Visit us at colecraftbooks.com or e-mail us at: colecraftbooks@embarqmail.com
Wholesale orders may be placed with our distributing partner, Ingrams